SpringerBriefs in History of Science and Technology

More information about this series at http://www.springer.com/series/10085

Leo Corry · Raya Leviathan

WEIZAC: An Israeli Pioneering Adventure in Electronic Computing (1945–1963)

 Springer

Leo Corry
The Lester and Sally Entin Faculty
of Humanities
Tel Aviv University
Tel Aviv, Israel

Raya Leviathan
Cohn Institute for History and Philosophy
of Science, Tel Aviv University
Tel Aviv, Israel

ISSN 2211-4564   ISSN 2211-4572   (electronic)
SpringerBriefs in History of Science and Technology
ISBN 978-3-030-25733-0   ISBN 978-3-030-25734-7   (eBook)
https://doi.org/10.1007/978-3-030-25734-7

This Springer imprint is published by the registered company Springer Nature Switzerland AG
The registered company address is: Gewerbestrasse 11, 6330 Cham, Switzerland

*This book is dedicated to Professor
Aviezri Fraenkel, תלמיד חכם, mathematician,
pioneer and adventurer in electronic
computing in Israel*

Preface and Acknowledgements

The research that led to the publication of this book was initially conducted by Raya Leviathan in the framework of her Ph.D. studies at the Cohn Institute of History and Philosophy of Science and Ideas, Tel Aviv University, under the supervision of Leo Corry. The dissertation, entitled "One State, One Computer. Building the WEIZAC Computer at the Weizmann Institute in the Mid-Fifties, the Decision, and its Effects," was successfully completed and approved in 2014. The central topic of the dissertation was the story of WEIZAC, the first electronic computer built and operated in Israel between 1954 and 1963 in the framework of the Weizmann Institute of Science (WIS) at Rehovot. Stories about WEIZAC have been, to be sure, far from being utterly unknown for many years now. On the contrary, personal reminiscences and anecdotal stories are found in many existing accounts of the activities of WIS and of the early years of the computing community in Israel, and they continue to surface occasionally in the Israeli media. However, prior to Raya Leviathan's Ph.D. dissertation, no systematic historical account had ever been undertaken, nor had the relevant historical records concerning the WEIZAC project ever been closely examined. Her dissertation shed new light on the unlikelihood of the project as it was initially conceived, as well as on the actual impact it had upon scientific activity in Israel at large and upon the rise of a local computing community. These important issues had never before been spelled out in detail.

The wealth of interesting documents and historical insights that arose from Leviathan's initial study were enlightening, surprising and promising. We thus decide to pursue additional research in this topic as a common undertaking of the two of us. We see this book as just the first one in a series of publications that we intend to undertake on topics related to the early years of electronic computing in Israel. These will be based on material that we have already gathered and studied as well as on additional material that we expect to collect and work out in detail in the near future. The topics we intend to consider include the following:

- The WEIZAC early computer as an accelerator to science in Israel.
- The development of computer science in WIS and other academic institutes in Israel. Was early entrance an advantage?
- Women in the early years of computing in Israel.

Work on this book has involved large amounts of archival research and personal interviews. In those cases where the archives provide keys to the documents, we have referred to the original archival key. We have also used documents that are undated or unsigned. In those cases, we simply cite by giving the available information. All the documents and photographs are cited and used by permission of the archives or holders in personal collections, as indicated in each case. We would like to thank and acknowledge those individuals and institutions, which have positively reacted to our enquiries, providing us with important information and insights, as well as permission to publish documents and photographs. In particular, we acknowledge and thank Anat Rotem-Braun, for her cooperation and permission to publish photographs from Werner Braun's archive.

We thank the Weizmann Institute of Science Archives (WIA), Rehovot, Israel, for their enthusiastic cooperation. At the Department of Applied Mathematics (DAM) of WIS, we benefited from the kind help of its Head, Prof. David Peleg, as well as of Administrative Assistants, Mr. Raanan Michael and Mrs. Carol Weintraub. They provided us with crucial access to the Chaim Pekeris Archive (CPA) and to the unpublished scripts of conversations of the department pioneers. The Weizmann Memorial Organization, "Yad Chaim Weizmann" (YCW), at Rehovot, provided us with access to and important guidance at the Chaim Weizmann Archive (CWA).

We want to express our sincere and special thanks to Aviezri Fraenkel, a prominent WEIZAC pioneer, for many informative and illuminating conversations. Likewise, we thank the late Gerald Estrin for sharing important information via email interchanges. Heartfelt thanks go to Ruth Riesel and Naomi Yaron, daughters of the late Zvi Riesel—another WEIZAC pioneer—for useful conversations and photographs taken from their personal collections.

Estrin and Fraenkel provided access to their private copies of six DVDs with enlightening video interviews conducted with the WEIZAC pioneers at WIS in April 1983. Now, the videos are freely available on a Web page entitled "The Computer Pioneers: Weizmann Institute Video Oral History." https://ethw.org/Archives:The_Computer_Pioneers:_Weizmann_Institute_Video_Oral_History.

Motti Golani and Jehuda Reinharz were kind enough to read and comment on parts of our manuscript while also sharing with us the final draft of their forthcoming biography of Chaim Weizmann. Their work presents a fresh, authoritative and comprehensive view of the life and work of this unique historical figure, and it helped us greatly in understanding his role as leader of WIS. We thank them for their collegiality and for sharing their insights.

For their comments and advice on preparing the final version of this book, we would like to thank Matteo Valleriani, Editor of SpringerBriefs in the History of Science and Technology, and Lucy Fleet and her team at Springer. We also acknowledge the helpful comments by an anonymous referee on a previous version of the manuscript.

This research was supported by The Israel Science Foundation (grant No. 279/16).

We are happy to thank our families for their continued support and for the sympathy and appreciation that they always express toward our work.

Tel Aviv   Raya Leviathan
March 2019   Leo Corry

The original version of the book was revised: Acknowledgement has been included and the missed author corrections in chapters 2 and 3 and the belated back matter corrections have been updated. The correction to the book is available at https://doi.org/10.1007/978-3-030-25734-7_5

Contents

Commonly Used Abbreviations and Acronyms

CPA   Chaim Pekeris Archive (DAM—WIS)

CWA   Chaim Weizmann Archive (The Weizmann House, Rehovot)

DAM   Department of Applied Mathematics

DBGA   David Ben-Gurion Archive (The Ben-Gurion Institute for the Study of Israel and Zionism on the campus of Ben-Gurion University, Sde Boqer)

HMF   History of the Mathematics Faculty, WIS. Unpublished scripts of conversations of Lee Segel with the DAM pioneers: Chaim L. Pakeris (February 24, 1987); Joe Gillis (February 10, 1987); Zvi Riesel (February 18, 1987) (kept at DAM secretariat)

HUA   Hebrew University Archive, Jerusalem

LPCW   The Letters and Papers of Chaim Weizmann (Yad Chaim Weizmann Institute, Rehovot). A partial selection of the letters and papers are reprinted in twenty-three volumes *The Letters and Papers of Chaim Weizmann: Series A (Letters)*, London: Oxford University Press; *The Letters and Papers of Chaim Weizmann: Series B (Papers)*, Transaction Publishers. Documents referred to above are cited in the footnotes under LPCW

RADC   Rafael Advanced Defense Systems—Business and Technology Information Center, Haifa

SCM   Minutes of the Scientific Committee Meeting, WIS

SIA   State of Israel Archive, Jerusalem

SOVA   Smithsonian Online Virtual Archive—Computer Oral History Collection

WIA   Weizmann Institute Archive, Rehovot

WIS   Weizmann Institute of Science, Rehovot

Chapter 1
Introduction

The State of Israel was just born and the small economy of the newly created country emerged in shambles from the war that had just ended. The Jewish community in Palestine, which in 1948, after the termination of the British mandate, numbered about 650,000, had a heavy defense burden as well as the burden of absorbing hundreds of thousands of refugees and immigrants. Hundreds of thousands of Palestinian Arabs had fled or had been forced to leave. The Jewish population in Israel doubled itself in the first three year of its independence. Israel faced problems that characterize young countries, but also problems unique to a country that absorbs masses of immigrants and struggles with a hostile political environment: economic, social, cultural and security problems, alongside the need to establish a civil rule of law. Food and foreign currency were lacking and the government implemented a regime of austerity, which was officially in place until 1959, and in which basic goods (including food, fuel, clothes and furniture) were strictly rationed.

Shaped by ideas and processes that originated very far away from the dramatic events that dominated post-war Middle East, this was also the dawn of the digital age and the rise of the new electronic computing technologies. These two separate and deeply unrelated historical threads, the creation of the State of Israel and the rise of the electronic computer, had a lasting impact, each in its own manner, on the way that the world would develop in the second half of the twentieth century. But curiously enough, their individual paths crossed with each other in a quite unlikely, yet highly important encounter in the 1950s. This happened when one of the earliest high-speed electronic, digital stored-program computers, WEIZAC, was designed and built in the sleepy farming town of Rehovot, about 20 km south from Tel Aviv.

Rehovot had been established in 1890 by a group of Polish Jewish pioneers belonging to the earliest wave of Zionist immigration to Palestine. In 1948, it had a population of about 12,500 inhabitants, most of them successful citrus growers. An Agricultural Research Station was opened in Rehovot in 1932, and two years later, in 1934, the Sieff Institute for Chemical Research was established very close to it.

© The Author(s), under exclusive licence to Springer Nature Switzerland AG 2019
L. Corry and R. Leviathan, *WEIZAC: An Israeli Pioneering Adventure in Electronic Computing (1945–1963)*, SpringerBriefs in History of Science and Technology, https://doi.org/10.1007/978-3-030-25734-7_1

The Sieff Institute was later expanded to become the Weizmann Institute of Science (WIS), which was formally opened in 1949. The Weizmann Institute would raise to become a world-class research institution, whereas Rehovot, alongside it and now with a population of nearly 150,000, turned into a busy hub of agricultural and scientific research and of global high-tech companies. Electronic computers, needless to say, dominate all aspects of local life in town, as well as the scientific activity of its institutions. But when the idea of building an electronic computer in Rehovot started to be discussed in the late 1940s, it sounded not only as a foolish idea in the context of an impoverished and troubled society, but also as one that was basically unnecessary for conducting scientific research as broadly conceived then in Israel.

The driving force behind the WEIZAC project was Chaim Leib Pekeris (1908–1993) a Lithuanian-born Jew, who received his basic mathematical training in leading academic institutions in the USA. He became acquainted with electronic computers as early as 1943 and had the opportunity to develop his skills, hands-on, working with the most advanced machines that were operational at the time. An enthusiastic Zionist, Pekeris attributed primary importance to science and technology as fundamental tools for promoting economic development and modernization in the new Jewish state. After joining WIS in 1948, he started to pursue his project for building an electronic computer as a flagship initiative for attaining such aims.

WEIZAC was built in the years 1954–55 and worked in full capacity for almost a decade. Its designers and builders used cutting-edge technology and achieved the highest benchmarks of computing performance at the time. The computer was modeled after the famous machine of the Institute for Advanced Study (IAS) in Princeton, which operated since 1952. In fact, the chief engineer of the project was Gerald Estrin (1921–2012) who had actively participated in the IAS computer project. Mathematicians and scientists from the Weizmann Institute and from other research institutions in Israel, as well as members of other Israeli government organizations, used the computer to advance science in Israel and to spread the word of this new technology all over the country.

This book tells the story of the WEIZAC project and of its immediate contribution to creating a computer-savvy community of users within the scientific and industrial realms in Israel, as well as in preparing the road for adopting the computer as a main tool in government and security agencies. The first of the two main chapters of the book explain the background to the apparently exceptional decision to build a computer in the challenging environment of the newly created State of Israel. We start (Sect. 2.1) with a brief account of the development of electronic computers in general, in the years before and during the construction of WEIZAC. We explain the innovative aspects of EDVAC, at the Moore School of Electrical Engineering of the University of Pennsylvania, and of the IAS machine at Princeton, two of the earliest machines to successfully implement the stored-program architecture in 1951, and we briefly discuss their overall impact. We then present a comparison of the situation in Israel with those of Taiwan, Ireland and India. These were three countries that in the 1950s, like Israel, had incipient economies and had only started to develop, with moderate success, local traditional industries. We find this comparison important because electronic computers were introduced in those countries much later than in Israel, and none of them came up at the time with a similar idea of a project to build their own enhanced machine. The background account and the comparison with other countries helps providing the right setting within which to understand the peculiarity of the WEIZAC story.

We then move to discuss the second component in the background to the project, namely the creation of WIS in Rehovot. In Sect. 2.2 we focus on the figure of Chaim Weizmann, in his dual role of well-known scientist and Zionist leader. We discuss his conceptions about the role of science as part of the Zionist project, and his somewhat ambivalent attitude towards the question of what kind of research, pure or applied, should be pursued in the newly created institutions of higher learning in Mandatory Palestine. Section 2.3 is devoted to the creation of WIS, initially as the Sieff Institute in Rehovot, and then in 1949 as the Weizmann Institute of Science. This section allows us to introduce the main players in the decision-making processes, and those who would have to approve and create the material conditions for carrying out the WEIZAC project. Section 2.4 focuses on the figure of Chaim Leib Pekeris, and on his views about computation-intensive science. We discuss here the question how these views contrasted with those pursued at the time at the Hebrew University in Jerusalem. Pekeris's views became fundamental for the kind of activities promoted at WIS and for turning the electronic computer into the tool without which no cutting-edge research could be carried out.

Against the background provided in Chap. 2, the second main chapter of the book discusses the processes leading to the decision to build the computer (Sect. 3.1), and the actual stages in its design and construction. This comprised several tasks that, considering the historical and geographical circumstances, can only be described as extremely unlikely: putting together a team of personnel equipped with the necessary skills (Sect. 3.2); building a lab and purchasing the required electronic components (Sect. 3.3); and in particular acquiring a "Magnetic Core Memory," the crucial component of which at the time existed only a handful in the entire world (Sect. 3.4).

Chapter 3 closes with a brief, but thoroughly documented analysis of the impact of WEIZAC on actual scientific research in Israel and beyond. This account makes clear the astounding extent to which research based on calculations performed with WEIZAC (as well as with the two machines that followed it at WIS, GOLEM and GOLEM B) were at the heart of the processes that turned WIS into the kind of world-class leading institution that it became. It also makes clear the extent to which WEIZAC was at the heart of the creation of a relevant community of scientists, engineers, technicians, and users, at all levels, of computing technologies in Israel, in its research institutions and in it government branches.

The concluding chapter summarizes the entire discussion, and locates it within the realm of broader historical issues, such as the role of science and technology in the process of nation-building in general and in the case of the State of Israel in particular. A main issue that arises in this context, and that appears as a connecting thread throughout the book, is the fundamental role played in this story by Jews and Jewish institutions all around the world, but particularly in the UK and in the USA. This is true for the people directly involved in the WEIZAC project in Rehovot, like Pekeris and Philip Rabinowitz, who joined WIS, above all, out of Zionist motivations. This is also true concerning the ways in which the Jewish institutions and the Jewish networks of cooperation worked on behalf of WIS and of the project, and without which the project could not have been successfully completed. The integration of the organizational mission as shaped by Weizmann himself and by the leaders of WIS, with the Zionist motivations of the key figures involved in the project, together with their views on the role of science as a main tool for nation-building, and with the aims of those who supported the project from abroad, were instrumental in leading to its eventual success. This is why it is fair to speak about WEIZAC not just as the first electronic computer to be built in Israel, but more specifically as the first Zionist electronic computer.

Chapter 2
Creating a Top-Rated Scientific Institution in Rehovot at the Dawn of the Digital Age

2.1 EDVAC and the Stored-Program Architecture—The Early Years of Electronic Computing

The story of WEIZAC is the unlikely story of how an electronic automatic computing machine was built and became operational during the early years of the State of Israel. Thus, the first context that needs to be clarified in order to provide a proper account of this story is the context of the early years of electronic computing in general and, more specifically, the period between 1945 and 1960. The present section provides a cursory background account of that remarkable and influential period in the history of contemporary technology, and of the place of WEIZAC within that context.

The history of automatic calculating machines is long and complex and it can be told in several ways, depending on the kind of emphasis laid upon specific features that relate to either their architecture, or their programmability, or their physical components, or several other possible perspectives (e.g. Mahoney and Haigh 2011). For the purposes of our account here, it definitely makes sense to start from the EDVAC as a most prominent milestone and starting point of the story. The main reason is that the EDVAC incorporated many crucial features that bring it closer than any other previous machine to the current conception of what is an automatic computing device.

These features were described in The *First Draft of a Report on the EDVAC,* a technical document published in 1945 by the brilliant mathematician John von Neumann (1903–1957), and which became one of the most ground-breaking and influential documents in the history of the modern electronic computer (von Neumann 1945). The ideas described in this document were developed in collaboration with the creators of an earlier machine, the ENIAC, since EDVAC was specifically planned as the ENIAC successor (Haigh 2016, Kindle Locations 505–506). The *Draft* put forward the basic logical design of a machine that was eventually

The original version of this chapter was revised: The correction to this chapter is available at https://doi.org/10.1007/978-3-030-25734-7_5

built at the Moore School of Electrical Engineering at the University of Pennsylvania, the "Electronic Discrete Variable Automatic Computer," EDVAC. It first became operational in 1951 and continued to run until 1962.[1] The importance of the *Draft*, however, lay in the concepts and approach put forward in it rather than in the specific details of the design presented (Priestley 2011, 142).

Although as actually implemented it differed from von Neuman's original design and architecture, the EDVAC was a high-speed, automatic calculator built purely out of electronic components, as opposed to some earlier machines that included also electromechanical ones. It was digital, as opposed to analog, fully automated as opposed to requiring manual intervention, and general-purpose, as opposed to special-purpose. Once these principles were simultaneously implemented within a single machine, the EDVAC, very soon they became *the defining features* of, and indeed, synonymous with, the idea of what a computing machine is. Interesting evidence for this is found in the proceedings of the earliest conference on electronic digital computers held in England, for which there are published records. The conference was held at the Royal Society on March 4, 1948, and the opening statement came Max Newman (1897–1984), one of the pioneers of the British electronic computing community.[2] He explicitly said that the discussion would be confined to machines that implemented all the main ideas that appeared in the EDVAC, namely:

> ...machines which are *automatic*, i.e. require no human intervention at any stage; *digital*, i.e. such that separate digits of each number are stored in the machine at every stage (in contrast to 'analogue' machines such as the Differential Analyzer where the numbers are represented by directly measured physical quantities, e.g. length); and *general-purpose* machine, i.e. machines able without modification to carry out any of a wide variety of computing jobs (Hartree et al. 1948, 265).

In addition to its architectural principles, in terms of hardware, the EDVAC incorporated the so-called "Delay line memory," the most advanced technology known at the time, with a capacity of 1,000 44-bit words. Its average addition time was 864 microseconds and its average multiplication time was 2,900 μs (Godfrey and Hendry 1993).

But without any doubt, the most important feature of the EDVAC was its being a "stored-program" machine: machine commands were fed into it via a separate input-output unit, and these commands were stored in the internal memory of the machine in a way that was essentially similar to that in which the data to be

---

[1]As an interesting anecdote we can point out here, that the chief engineer of the two GOLEM projects (see below), Smil Ruhman (1925–) joined WIS in 1961 after having worked at the EDVAC as a student at the Moore school. His task in that project was to document the design. According to his testimony (Interview with Ruhman, by R. Leviathan in Dec. 11, 2013), this was the best way to learn about the design of electronic computers.

[2]Maxwell Newman led one of the Bletchley Park groups during World War II. After the war, he participated in the Mark 1 project at the University of Manchester. See (Computer Pioneers— Maxwell (Max) Herman Alexander Newman [Changed by deed poll in 1916 from Neumann.] 1995), https://history.computer.org/pioneers/newman-mha.html (accessed Jul. 7, 2018).

operated upon was stored.[3] It is generally accepted that this specific feature implied a most crucial breakthrough in the design of automatic computing machines (Haigh et al. 2014). To this day, it has remained as a most essential characteristic of general-purpose electronic calculating devices and, in historical perspective, it was certainly the key ingredient leading to the seminal separation between software and hardware as two different, if intimately connected, aspects in the conceptualization, design, implementation and utilization of computers.

In the summer of 1946, as a result of the great interest aroused by the new automatic computing machines, the Moore school organized a course that introduced the basic concepts in this new field of knowledge, paying particular attention to an in-depth acquaintance with the stored-program architecture (Campbell-Kelly and Aspray 2004). This course, *Theory and Techniques for Design of Electronic Digital Computers,* also known as the "Moore School Lectures," became a most influential landmark in the spread of knowledge about the new technology of electronic computing (Haigh 2016, Kindle location 4357–4361). The list of lecturers featured the most prominent names in the field at the time, and of course von Neumann himself.

Von Neumann was the driving force behind another highly important computer project, conducted at the Institute of Advanced Study (IAS) at Princeton. The IAS machine was built between 1945 and 1951, with Julian Bigelow (1913–2003) as its leading engineer. Operational until July 1958 (Dyson 2012, 318), it embodied many of the principles that von Neumann had first introduced in his *Draft* of 1945, but also comprised several significant improvements that he introduced in subsequent work (Bigelow 1980; Burks et al. 1946). Among these was the transition from a serial arithmetic processing unit to a parallel one. Full operational implementations of the "storedprogram" idea could be found in machines built somewhat earlier, such as Cambridge EDSAC or the Manchester MARK I (Campbell-Kelly and Aspray 2004), but none of them was as influential as the IAS machine, in terms of the number of similarly designed machines built all around the world. The importance of the IAS machine project was manifest not only in terms of its logical design and organization, but also in terms of the physical components that were used in its construction and the engineering procedures developed while it was being built and initially put to work.

The basic *logical* principles behind the "von Neumann Architecture" are still influential in the current design of computers, but the *physical embodiment* of this basic design has changed enormously. The transition from vacuum tubes to semiconductors is among the main noticeable changes. More generally, all what concerns size, speed of the processing units and the capacity and efficiency of memory and storage devices has undergone sea changes. In order to understand the orders of magnitude of these parameters as they appeared in what was at the time a

---

[3]In the original EDVAC draft, each memory word included a special bit to indicate if it represented data or a command.

Table 2.1 IAS computer versus portable PC—a comparison of technical specifications

	IAS computer	Portable PC (Lenovo Yoga + Intel® Core™2 Processor)	Factor of change
Physical size	3.0m × 3.2m × 8.4m (room size)	0.4m × 0.02m × 0.3m, (processor size —37.5mm × 37.5mm)	3×10^{-5}
Memory	1,024	8,000,000,000	8×10^{6}
Word size	40	64	1.6
Number of components	2300 vacuum tubes	410 Million transistors	1.8×10^{5}
Instructions per second	~ 10,000	~ 2,500,000,000	2.5×10^{5}
Price	650,000$	1000$–500$	650

groundbreaking machine, it may be illustrative to compare the technical specifications of the IAS machine with a present-day device (Table 2.1).[4]

Possibly under the decisive influence of von Neumann himself, the technical documentation of the IAS machine and the ideas related to it were widely circulated, and hence the project became immensely influential in the development of the electronic computer industry worldwide (Aspray 1990, 91–94). Copies of the IAS machine soon started to appear beginning in the 1950s and up to the early 1960s, both in the USA and abroad. Among the best-known examples of these early machines (and the respective years when they became operational), the following can be mentioned[5]:

- PERM in Munich (1950);
- ORDVAC at Aberdeen Proving Grounds (1951);
- ILLIAC at the University of Illinois (1952);
- MANIAC at Los Alamos Scientific Laboratory (1952);
- AVIDAC at Argonne National Laboratory (1953);
- ORACLE at Oak Ridge National Laboratory (1953);
- BESK in Stockholm (1953);
- BESM in Moscow (1953);
- DASK in Denmark (1953);
- JOHNNIAC at RAND Corporation (1954);
- SILLIAC in Sydney (1956).

---

[4]Compiled from Intel Microprocessor Quick Reference Guide—Product Family, https://www.intel.com/pressroom/kits/quickreffam.htm#core2 (accessed May 30, 2018); (Estrin 1952).

[5]Information compiled from various sources: Aspray (1986), History| Argonne National Laboratory https://www.anl.gov/about-argonne/history; *Digital Computer Newsletter*, Office of Naval Research—Mathematical Sciences Division (Vol. 5 (1 and 4); Vol. 6 (1 and 2); Vol. 7 (3); Vol. 11 (3); *Electronic Computer Project, IAS,* 2017. https://www.ias.edu/electronic-computer-project (accessed Dec. 18, 2017). See also Prokhorov (1999).

WEIZAC was part of this trend, and even its name—with the ending AC standing for "Automatic Computer"—followed what became an accepted convention in the field.

By 1950 there were about ten automatic computing research centers around the globe. Less than one thousand people were seriously interested in this new technology at the time, most of them in the US and Great Britain (Aspray 1985, ix). Still, within a few years, one dozen countries had computers, either working or under construction. Not all of them ever reached full operation and only part of them were electronic digital stored-program computers per se. In a survey on Automatic Digital Computers, based on information gathered in the beginning of 1953, and conducted by the Office of Naval Research, we find a list of nearly one hundred such machines, nationally distributed as follows: USA (~ 70), Britain (10), Germany (8), Japan (3), Canada (2), France (2), Holland (2), Sweden (2), Swiss (2), Belgium (1), Australia (1), Norway (1) (Blachman 1953). In 1955 there were already about 200 computers in about 15 countries (Aspray 1986).

Only developed countries could afford at that time the necessary resources and the appropriate technologically advanced industry needed for building and running such automatic computing machines (Cortada 2013). Obviously, the budding State of Israel did not belong to this league. Its economy was of a much smaller scale, and it continued to suffer from the traumatic impact of the recently finished war. In its early years, the Jewish population doubled itself in three years, and the Israeli society faced enormous challenges that involved defense as well as economic issues, together with the burden of absorbing a huge amount of immigrants and refugees.

Thus, in order to grasp the true import of the WEIZAC project and of its impact, it is convenient to compare the Israel computing scene in the 1950s with that of countries like Ireland, Taiwan and India, rather than with those of the more developed ones mentioned above. In the 1950s, Ireland, Taiwan, and Israel were three comparable countries in the sense of being poor, peripheral, and technologically backward, with moderate success in traditional industries (Breznitz 2007, 6). However, in the late 1960s, all three had initiated local, tech-savvy industries led by the state. Then, in the 1990s, all of them experienced a marked wave of industrial innovation, to the point of becoming prominent players in the global market. India achieved independence in the late 1940s, about the same time as Israel, after a period of British rule. India and Ireland, like Israel, are all well-known for the remarkable growth of their software industry at the end of the twentieth century (Arora and Gambardella 2006). It is thus relevant to conclude this section by taking a closer comparative look at the early history of computing in these countries.

The first automatic computing equipment arrived in Taiwan in 1959. It was IBM equipment for punching cards. It was used for the purposes of accounting in contexts such as the US Aid Project, the population census, and the creation of local economic statistic institutions. With the assistance of the United Nations, in 1962 a large computer, the IBM 650, was leased by National Chaio-Tung University (NCTU), and two years later came a second one, an IBM 1620 (Tinn 2010).

Ireland was ahead of Taiwan in using computers, but it did not take part in developing the first generation of electronic automatic calculators. The first computer arrived in Ireland in 1958 and it was installed at the Irish Sugar Company. It was a British computer, the ICT 1201, manufactured by the British Tabulating Machine Company (BTM). About three years later, in 1962, an IBM 1620 Model 1 was installed at the Trinity College of Engineering in Dublin, but access to electronic computing was limited to a small community. It was only in the late 1960s and early 1970s that some major American high-tech companies arrived in Ireland and that the Irish government, through the Industrial Development Authority (IDA), attracted IT investments (Sean 1997).

In India in 1955, as a first step towards the development of a full-size electronic computer, the electronics department of the Tata Institute of Fundamental Research in Bombay launched a project for the designing, and then actually building, an experimental computer modeled after the IAS machine. The machine, however, designed and built between 1955 and 1957, was never put to use in actual practical tasks, due to some significant technical limitations in its implementation.[6] Still, as a pilot model, it achieved its goals. By allowing the team's members to acquire experience and knowhow, as well as providing a platform for testing their new ideas, it gave its designers the confidence to move to a larger computer, with a 40-bit word and a magnetic core memory of 1024 words (Rao 2008). This second machine, TIFRAC, was built within three years and was completed in 1960. The machine had innovative features and it opened the way for setting up a National Computational Facility in India (Rao 2008, 428). However, this happened at a time when the "second generation" of digital computers, that implemented the new transistors technology had already entered the marketplace. Such were the Philco TRANSAC S–1000 and S–2000 machines (Ceruzzi 2003, 65). Later, the Indian Institute of Statistics and the University of Jadavpur in Calcutta constructed another computer, the ISIJU–1, which became fully operational only in 1966 (Banerjee 1996, 22).

The use of computers in business and industry in India began in 1961, when an IBM 1401 machine was installed at the ESSO local headquarters in Mumbai. This marked the beginning of the era of commercial computers in the country. During the 1960s, computers were installed in research and educational institutions. The best known were the CDC 3600, installed at the Tata Institute, and the IBM 7044 at the IIT (Indian Institutes of Technology) in Kanpur. In 1970 there were about 120 computers installed in India in various organizations (Achuthan et al. 1992, 139; Banerjee 1996).

The examples of these three countries provide the proper context for understanding the historical import of the WEIZAC project and its uniqueness: in the early years of the State of Israel, a state-of-the-art electronic computer was

---

[6]Specifically: the Tata computer had only 100–200 words in memory. Each word comprised 11 bits. The instruction set of the machine was very limited. See *Digital Computer Newsletter*, Office of Naval Research Vol. 9 (1).

Table 2.2 Early computers by country

	India	Taiwan	Ireland	Israel
Indigenous early computer	1957: Pilot model 1960: Full-fledged early computer	–	–	1955
First computer installation	1960	1962	1958 (business computer) 1962 (scientific computer)	1955

constructed at WIS, and it was adopted for significant use at other institutions. This happened no less than five years prior to a similar situation in any of the comparable countries (Table 2.2).

The decision to build an improved version of the IAS computer at WIS in the 1950s appears, under this perspective, as truly exceptional. The processes that started in 1946 and that eventually led to the decision to build the machine, and the way in which the project was effectively and successfully implemented at WIS in 1954–1955 become, indeed, a matter of distinct historical interest. In the following sections, we discuss the details of this remarkable process.

2.2   Chaim Weizmann: Reputable Scientist and Zionist Leader

The second context that needs to be clarified in the background to the story of WEIZAC is the institutional context, namely, that of the early years of WIS, and of the people who were involved in its creation. Foremost in this context is, of course, the figure of Chaim Weizmann (1874–1952) himself, in his dual role of reputable scientist and prominent Zionist leader. These two aspects of his life were not disconnected from each other. Indeed, science and technology played a significant role in Weizmann's Zionist vision, and he was actively involved in the founding of two main academic institutions in Palestine, the Hebrew University in Jerusalem and the Weizmann Institute in Rehovot. A most pressing question that occupied the mind of all those involved in the creation of these two institutions, not only Weizmann, concerned the way in which they should and would contribute to the broader aims of the Zionist project. This question translated into concrete debates about the disciplines that should be cultivated in them and about their orientation. A particular focus was on the question of the relative importance to be accorded to theoretical (or pure, or basic) versus applied branches of science. Weizmann's views in this regard underwent interesting changes and probably they were rather ambivalent all along the way. At any rate, they were highly influential on the way in which WEIZAC project developed.

In this section we want to discuss some important aspects of Weizmann's biography that are necessary for understanding the background to the creation of the institute in Rehovot, and the vision that underlie it.[7] Chaim Weizmann was born in 1874 in the town of Motol, near Pinsk, in Belarus. In 1892, he moved to Darmstadt and then in 1894 to Berlin, to study at the local *Technische Hochschule*. In Berlin he became involved in Zionist activities, together with other young students, such as Shmariyahu Levin (1867–1935) and Leo Motzkin (1867–1933), who, like Weizmann, had arrived in Germany from the territories of the Russian Empire and who would eventually become prominent Zionist leaders. Starting from the second one held in 1898 in Basel, Weizmann attended most of the Zionist Congresses held during his lifetime. With time, his own status and leadership position within the Zionist Organization consolidated and he was elected president in two different periods: 1920–1931 and 1935–1946. In 1904 he settled in the UK where he would remain for the next forty-five years, with some periods outside the country. It was only in 1949, as he became the first president of the State of Israel, that he renounced the British citizenship.

On the scientific track, Weizmann's career started in 1899, when he completed his doctorate in organic chemistry at the University of Fribourg, Switzerland. In 1901 he joined the Department of Organic Chemistry at the University of Geneva. In 1904, he received a job proposal from the Department of Chemistry at the University of Manchester in England, and he was appointed senior lecturer there. Eventually, however, he did not develop there a long-term, stable academic position. His working relations with the department were never truly smooth, and tensions kept rising around the issue of royalties deriving from Weizmann's inventions. These were years of great success for Weizmann as a Zionist leader, and he also continued to be involved, albeit to a lesser degree, in scientific activity. Right before World War I, Weizmann invented a method to produce synthetic acetone out of corn. The method, which facilitated the production of gunpowder, became a significant asset to the British war effort as well as a source of significant income for Weizmann himself. Weizmann's political wit helped him translate this achievement into both an increased recognition of his reputation as a scientist and an empowerment of his already existing access to the corridors of power in London. After 1915, while he continued to be involved in scientific activity to a much lesser degree, Weizmann's interest in Zionist politics became his main passion and it occupied most of his time and energies. It was at this time that he became a leading figure in the Zionist movement.

The idea of establishing an institution of higher learning in Palestine, meant to serve as a spiritual center for the Jewish people, became a prominent issue in the deliberations of the Zionist congresses since they first were convened in Basel in

---

[7]The research literature on Weizmann is rich and extensive and we rely here on that wealth of works. See, e.g. Cohen and Chazan (2016, 379–383, 421–434), Fischer (1994, 287–231), Golani and Reinharz (2019 (forthcoming)), Reinharz 1993, Rose (2015 [1987]), Weisgal (1971). Only in relation with certain points of particular interest, we provide below more specific references to this secondary literature.

1897. Weizmann eventually became one of the enthusiastic supporters of this idea, as did also Albert Einstein (1879–1955). Weizmann's beliefs about the primacy of culture and science as spiritual assets of inherent value deserving active promotion along with political activity were part of his own view of "synthetic Zionism." This was the view that he had first introduced in 1907 at the eighth World Zionist Congress, as a synthesis of the methods of the political Zionists who stressed the importance of political action and wanted a state and the practical Zionists who sought to build up a homeland through immigration, settlement, and institution-building (Ball et al. 1996, 531; Wagner 2015). He expressed from very early on the belief that this synthesis was the adequate way leading to the establishment of the national homeland for the Jewish people in Palestine, and in this context, science would become an important tool for shaping Jewish identity in the new national homeland.

As part of their academic and Zionists views, both Einstein and Weizmann explicitly opposed the idea of establishing a university with low academic level, meant to meet, above all, basic employment and subsistence needs of the Jewish population in Palestine. This was a topic of acerbic debate already in 1910, between Weizmann and the revisionist leader Ze'ev Jabotinsky (1880–1940) (Kolatt 1997, 18–19). High-level scientific activity seemed to Weizmann more important than the learning of practical skills. At the dedication ceremony of the Hebrew University, on April 1st, 1925, Weizmann was very explicit in expressing such views (Reinharz 1997, 320–326). He brought up the issue of scholarship as a prominent trait in the history of the Jewish people, and emphasized the high value attributed throughout the generations to literacy and study, regardless of economic and political status. Nonetheless, Weizmann also expressed his appreciation for Jewish institutions already functioning in Palestine, which were initially intended as vocational schools that would serve the agricultural and industrial sectors. These were the Institute for Agricultural Research established in Tel Aviv (and later moved to Rehovot, see below) and the Technion in Haifa.[8]

This significant ambivalence has been widely documented by way of testimonies of people who were in contact with Weizmann in various opportunities (Berenblum 1966; Sieff 1970). Of particular interest is the testimony provided by the journalist Ritchie Calder (1906–1982), who in 1959 wrote a book on the first decade of WIS (1959), and had lengthy conversations with Weizmann's closest collaborators. He thus wrote:

> When, as has happened periodically, there are debates as to whether the Weizmann Institute should undertake teaching or should turn to researches that show practical possibilities, to commercial purposes, Weizmann can be invoked both for and against. He did remove the Institute, in its origins, from proximity to the Hebrew University of which, as a teaching institution, he had laid the cornerstone on Mount Scopus, but he put it next door to the Agricultural Research Institute so that scientists would not forget that there were practical

---

[8]Weizmann, "Speech at the Opening Ceremony of the Hebrew University, Apr. 1, 1925". Quoted in Katz and Heyd (1997, 319–322).

applications as well. These debates have been resolved with the pragmatism he himself would have applied (Calder 1962, 122).

It seems that Weizmann was eager to promote simultaneously what could appear to be two opposing tendencies. On the hand, he was committed to the highest standards of scientific research, with their long-term implications for the Zionist project and the possibility of placing the Jewish national entity in a prominent place among the civilized peoples of the world, which he felt to be part of. On the other hand, he acknowledged the immediate needs of the Jewish settlers and immigrants and the obligation to contribute in that direction as well. His support for the WEIZAC project was strongly related to the former consideration—an electronic computer as a tool for cutting-edge scientific research—but also the development of high-level technological capabilities was in itself a strongly appealing motivation for him.

The Board of Governors of the Hebrew University in 1925 elected Weizmann as President, Judah Leib Magnes (1877–1948) as Chancellor, and Einstein as the Chairman of the Academic Council. Differences in their respective approaches to the desired aims of the university, the provision of undergraduate education and the need to conduct advance research, as well as confrontations around matters of authority, soon led to deep clashes among them. Initially, Magnes was responsible only for the university's finances and administration, but he aimed at extending his authority also to academic matters. Both Weizmann and Einstein, on their side, did not trust the "local" leadership and its ability to keep high academic standards. Magnes' conception was that the Chancellor is "the head of the whole university,"[9] whereas Weizmann objected to the subordination of the Academic Head to the Chancellor. In addition, political differences between Weizmann and Magnes became increasingly pressing, as the latter strongly advocated for a binational state in Palestine (Kotzin 2010, esp. Chap. 6). The question of the desired relation between the Zionist administration and the University also became a main issue of conflict between the two. Eventually, after ten years of heading the University, Magnes' leadership came to an end in 1935, and he accepted the honorary title of President of the institution, a title which he held until 1948 (Goren 1997).

Einstein's political views were closer to those of Magnes, whereas his academic views were closer to those of Weizmann. He kept a skeptical distance from all direct involvement, and he consistently presented himself as a "supporter of Zionism" rather than a "Zionist." He emphasized community bonds and cultural motivations that could seem less intrusive than actions to gain control of land and power. In the end, the significant differences with Magnes over academic policy and Einstein's disgust with power struggles were stronger than the political affinities between the two, and this soon led to Einstein's resignation from the Board already in 1928 (Corry and Schappacher 2010, 440–446; Katz 2004; Kirsh and Katzir 2016, 422). Einstein remained committed to the Hebrew University, but always from a prudent distance.

---

[9]Magnes to Weizmann, Oct. 24, 1928. Quoted in Parzen (1970, 187–213).

Both Einstein and Weizmann laid the stress on pure or basic science as an intellectual endeavor of great cultural value, but this is not to say that Weizmann did not acknowledge the usefulness of applied science to the healthy development of the Jewish Yishuv in Palestine.[10] As a scientist, Weizmann himself had personally benefited from his involvement in applied science and the sales of his inventions. From this perspective, the practical contribution that science and technology would provide to the residents of Mandatory Palestine and to its neighbors was an additional important aim for the Zionist project and one of the immediate tasks that it should undertake. Thus, for instance, in an interesting speech delivered in Tel Aviv in 1936 about the connection between Torah and day-to-day practical concerns he expressed this view in the following words[11]:

> In these days, I had the opportunity to visit several factories and look at the development of the industry, and I saw very important beginnings that already occupy a large part of the country's economy. In every organized country, there is a strong connection between science and industry. Today it is impossible to organize an industry that can compete with the international industry without a scientific basis. However, science is changing from day to day, because it too is progressing and industry must take into account the progress in science. The beginnings of our industry are beautiful, but they need connections with science (Weizmann 1936).

In the seventeenth Zionist Congress of 1931, Weizmann was not re-elected to the presidency of the movement. Weizmann saw this step as a de facto demotion from the post that was so important for him. Disappointed and offended, Weizmann started to look for new paths of activity, including a possible return, perhaps partially, to active scientific research (Fischer 1994, 288). A couple of months before the congress, Weizmann had requested to set up a laboratory at the Hebrew University in Jerusalem. It seems that actual scientific research was not the main motivation behind Weizmann's. A stable basis of activity for him in Jerusalem seems as a more realistic explanation. Indeed, in a letter to Einstein, he explained that presently he could not fulfill his desire to come to Jerusalem, what had always been his "ideal," because he did not want to hurt his son at school at his "tenderest age." However, he considered to spend one semester every year in Jerusalem, for the next three years, until his son would start attending university (see also Kirsh and Katzir 2016, 421).[12] One way or another, the Hebrew University rejected Weizmann's request on the pretext of lack of funds. It is likely, however, that the real reason was Weizmann's perceived ambitions to become the academic head of the University, which became a further matter of concern for Magnes.

Disappointed also on this front, Weizmann decided to set up a small private laboratory in London, where he could start conducting new chemical experiments. But once again, it seems that the continued pursuit of political aspirations within the

---

[10]The term "Yishuv" was an accepted term to refer to the Jewish community in Palestine, since the turn of the twentieth century and before the establishment of the State of Israel.

[11]Original in Hebrew. Unless otherwise stated, all translations from Hebrew are by the authors.

[12]Weizmann to Einstein, May 21, 1931 (LPCW Series A, Vol. 15, Letter 152).

Zionist movement remained Weizmann's main focus of interest and activity, clearly overshadowing his scientific concerns. Indeed, shortly after establishing this lab, he delegated its functioning to an assistant and travelled to South Africa to meet with local Zionist activists (Golani and Reinharz 2019, Chap. 13). Nonetheless, this was also the time when the process that eventually led to establishing a new research institute in Rehovot, as well as his own home in the same time, started to take shape.

2.3 Creating an Institute in Rehovot (1934–1949)

The cornerstone-laying ceremony of the "Daniel Sieff Institute for Chemical Research" in Rehovot took place on April 1933, and the opening ceremony on April 1934. Weizmann's project for a new research institute in Mandatory Palestine materialized rapidly thanks to the active support of his circle of Zionists friends in Manchester, sometimes referred to as the "Manchester School."[13] Of fundamental importance in this regard was the generous endowment established by wealthy Sieff family, led by Israel Sieff (1889–1972) (Cohen and Chazan 2016, 377–411; Fischer 1994, 289; Golani and Reinharz 2019, Chap. 14; Sieff 1970, 157). Indeed, the institute was dedicated to the memory of Daniel Sieff, late son of Weizmann's old friend, who had passed away on tragic circumstances.

There were several reasons for Weizmann's choice of Rehovot as the location for his planned new house and his new institute (Cohen 2016), but it is important to stress, in the first place, that the creation of the institute was not the result of a well-conceived plan, clearly established in advance and systematically developed thereafter. Rather, the idea evolved gradually within changing historical and personal circumstances, until it achieved its full-fledged format, and eventually materialized the way it did. Indeed, in the wake of their confrontation with Weizmann, the leaders of the Hebrew University came up with an initiative to compensate him by way of an offer to establish a research facility or, more precisely, a faculty for research in the field of agriculture, whose scientific leadership would be in the hands of Weizmann. The year was 1932, and that time, the Agricultural Experiment Station, initially founded at Tel Aviv by Yitzhak Elazari-Volcani (Wilkanski) (1880–1955) with the support of the Jewish Agency, had moved to Rehovot. As already indicated, Weizmann had mentioned the station on various occasions (Jensen et al. 2011, 40), and he eventually proposed cooperation between the station and the planned research facility (Cohen 2016; Weizmann 2013, Kindle Locations 3038–3040). Weizmann thought that Elazari-Volcani would be a reliable scientific ally for him and was eager to develop their collaboration.

---

[13]See, for example, www.manchesterjewishstudies.org/manchester-school/ (Retrieved Feb. 22, 2019).

The idea of the research institute started to take shape, and in its initial conception, its goals were rather circumscribed and even modest: to develop local and imported plant species on behalf of the pharmaceutical and chemical industry (Cohen 2016, 64; Rose 2015, Kindle Location 6210). Weizmann referred to this important issue in his inaugural speech, where he highlighted once again the two leading threads of his own public persona, Zionism and science, as well as the dual role, theoretical and applied, that he envisioned for the kind of knowledge that he expected to help achieve:

> The new institute will primarily devote itself to pure scientific research in all fields of agricultural and biological chemistry. It hopes, at the same time, through the application of modern scientific methods to the actual problems which face agriculture in this country, to contribute to the development of Palestinian agriculture, and so, directly or indirectly, to help in obtaining from the soil of Palestine products and commodities which, by their quality, are capable of holding their own, and more than their own, in the markets of the world... Room will also be found for promising students, who, after finishing their formal course at the university, wish to specialize in this particular branch of their subject.[14]

The very possibility of establishing a new center of research *outside Jerusalem*, certainly added much weight in favor of Rehovot, from Weizmann's personal perspective. As a matter of fact, Weizmann was intent on cultivating an alternative public image of the scientific community that he would help create now. This image would shy away from what many prominent voices in the Yishuv tended to associate with the Hebrew University and with its faculty members: a spirit of seclusion and overall estrangement of the immediate aims of the Zionist movement.[15] Weizmann aspired to establish a model of national science that would connect with the universal values of pluralism, excellence and innovation, but which, at the same time, would have a significant impact on attracting, and helping absorb Jewish scientists intent on immigrating to Palestine following the call of their Zionist beliefs, or escaping from anti-Semitic hostility anywhere around the world (Cohen 2016, 17). The dramatic events in Europe and the rise to power of Hitler in Germany and Einstein's announcement that he would not return to Germany had a great impact on Weizmann, and they left a mark on Weizmann's focus of activities.

---

[14]Weizmann, "Speech at the Opening Ceremony of Daniel Sieff Research Institute, Rehovot, Apr. 3, 1934" (LPCW, Series B, Vol. 2, Doc. 10.).

[15]As a matter of fact, some leading figures in the Hebrew University held political views that were far apart from the mainstream Zionist movement in Palestine, concerning the future of the relations between Jews and the local Arab population. Many of them were associated with the small but well-known Brit Shalom ("covenant of peace") movement, that supported the cause of a peaceful coexistence between Arabs and Jews, at the price of Zionism renouncing the wish to create a Jewish state. The group included, among others, Magnes, the Kabbala researcher Gerschom Shalom (1897–1982), the philosophers Martin Buber (1878–1965) and Shmuel Hugo Bergmann (1883–1675), and the educator Akiva Ernst Simon (1900–1988). See Keidar (1976), Kotzin (2010). Einstein consistently expressed his sympathies for the ideas of this group. See Rosenkranz (2011, 215–227).

The role he devised for a new institute in Rehovot as a magnet for encouraging the immigration of talented Jewish scientists became prominent in Weizmann's agenda. According to an often-told story about the visit in 1936 of Earl William Peel (1867–1937), Chairman of the British Royal Commission on Palestine, to the Sieff Institute, the visitor found Weizmann at work in his laboratory and asked him about his doings. "I am creating absorptive capacity," was Weizmann's reported reply.[16]

Eventually, following the Holocaust, with the frightful spiritual and intellectual losses sustained by the Jewish people, the issue of immigration became even more pressing for Weizmann. He stressed the point in his autobiographical book *Trial and Error*, in the following terms:

> The creation of scientific institutions in Palestine is essential if we are to insure the intellectual survival of the Jewish people. It may take us as much as fifty years to regain our strength in this field, and the only hope is that the men of high qualification who come to us will influence the young generation of Palestine in the direction of skill, discipline, order, and high quality performance (Weizmann 2013, Kindle Locations 5332–5335).

Well-established European scientists, however, did not really join the institute in its early years. Perhaps this fact turned out to be an important reason for the eventual success of WIS. Instead, the institute did attract many young scientists who had only started to build careers and followed newly opened directions of research that turned out to be in the cutting edge of international science in various fields of knowledge. As will be seen below, in the case of the WEIZAC, the ability to attract promising Jewish young scientists to Rehovot, such as Pekeris, became a key factor in the eventual success of the project.

About a decade after its foundation, Weizmann's friends in the US decided to enlarge the scope and goals of the Sieff Research Institute and to expand it into a full-fledged research campus, as a tribute to Weizmann on his seventieth birthday. The new institute was renamed the "Weizmann Institute of Science." The American Committee for WIS was established on October 1944, and it immediately initiated its fund-raising activity. A Scientific Planning Committee undertook the tasks of defining research plans, designing and constructing new buildings and equipping the new research institute.[17]

Two individuals played key roles in this process. First was Ernest David Bergmann, (1903–1975), the Scientific Manager of the Sieff Research Institute. Bergmann had earned his Ph.D. degree in organic chemistry in 1927 at the University of Berlin, where he also became a lecturer. In 1933, after the Nazis came to power, he made the decision to leave Germany and joined Weizmann in his private laboratory in London. Weizmann proposed to employ him as the head of the Sieff institute, then under construction in Rehovot (Jensen et al. 2011). After the

---

[16]Weisgal, "Report by the Chairman of the Executive Council for the Period from November 2, 1949 to Jun. 30, 1952 (draft)" (WIA 12–90–17).

[17]Weisgal, "Report to the Board of Directors of the American Committee and to Committees in Other Countries for the Weizmann Institute for the Period of 1944–1949" (WIA).

announcement about the intended expansion, Bergman devoted time and efforts to the planning of the new institute, of which he eventually became scientific director in its first years.

Bergmann and Weizmann were in close relationship and their partnership was a key ingredient in the early success of WIS. Like Weizmann, also Bergmann was fond of stressing the view of WIS as the harbinger of a new attitude on the side of local scientific institutions in Palestine, actively mobilized on behalf of the aims of the Zionist project. In an article published at the occasion of the tenth anniversary of the creation of the Sieff Institute in the daily newspaper of the Labor movement, under the suggestive title of "A Decade of Scientific Pioneering," he drove this message home very clearly when he wrote:

> Over the last years, much debate has been held around the question whether science should be pursued as an abstract endeavor, which keeps its distance from the pressing needs of the day and refrains from influencing real life, or if it should be fully devoted to improving human life. This question was definitely solved at the Sieff Institute from its inception ten years ago: the research methods that dominate its activities have to be fully objective – but its right to exist is directly based on recognizing the deep subjectivity of its leading aim: to turn the Land of Israel into the Homeland and shelter place for the masses of suffering Jews from around the world, and to revive it as a center of culture and inspiration for all peoples.[18]

Still, significant differences between the two arose along the years, especially in matters related to the scientific orientation of the institute and to the interrelation between the institute and the concrete needs of the Zionist movement and later, the newly created state. We can appreciate this point, for instance, in a letter to Bergmann in 1946, where Weizmann returns once again to the issue of practical versus pure goals of the new research institute:

> I think that in the future we should leave off the technical research, and pay attention to purely scientific work… You speak of the necessity of contributing some money to the research on technical problems… On no account would I like to have technical research in which I am personally interested carried out in Rehovot. It would be a very wrong policy, and would lay us open to great deal of misunderstanding.[19]

By the beginning of the 1950s, an impasse broke out between the two, as Bergmann grew closer to Ben-Gurion, Weizmann's fiercest rival in the Zionist movement. Bergmann became involved in secret military projects, and at the same time he also became "increasingly obsessed with the lack of military preparedness on the part of the Jewish community" (Jensen et al. 159). Deep differences of opinion arose between Weizmann and Bergmann over the relationships between the scientific activities at the institute and the security needs of the state (Cohen and Chazan 2016, 217–218).

In 1951 Bergmann left WIS. He taught some chemistry courses at the Hebrew University and was appointed personal advisor to Ben-Gurion. Beginning in 1952

---

[18]Bergmann, A Decade of Scientific Pioneering. *DAVAR* Nov. 11, 1944 (Hebrew).

[19]Weizmann to Bergman, Oct. 8, 1946 (CWA 18–2695).

he became one of the central figures of the new Israel Atomic Energy Commission (Deichmann and Travis 2004, 59). By 1958 he had already adopted a much more extreme view about what he saw as the correct relationship between science and the state. In a speech about the achievements of Israeli science, he said:

> Scientists must know that science is no longer a goal for itself, but a means to strengthen the state's position and ensure its development and security.[20]

The second important player in the creation of WIS, especially in matters related to the gathering of financial support, was Weizmann's personal assistant from 1940, Meyer Wolfe Weisgal (1894–1977). Weisgal emigrated at the age of eleven with his parents from Poland to the United States. He graduated in 1916 with honors from Columbia University (Tidhar 1959). During the 1930s he worked in the theater business, and was involved in the production of several plays. In 1937, one of his productions was the "The Ethernal Road", a cautionary tale aimed at warning against the persecution of Jews in Nazi Germany. Weisgal was active in the Zionist Organization of America, and in 1940 he became Weizmann's personal assistant in the US. He then moved to the recently created State of Israel, and became Chairman of the Executive Committee of WIS. Weisgal soon enlisted the help of his many friends in the United States on behalf of the institute (Weisgal 1971). He was a most influential figure in WIS until the day of his death in 1977.

As World War II ended and the dimensions of the Holocaust were exposed, the American Jewish community called up to raise rehabilitation funds for the survivors. Consequently, some voices urged to delay the ambitious plan of enlarging the Sieff Institute in face of the political disorder and violence that broke out in Palestine following the military activity of the Jewish resistance groups against the British Rule. In his report as Chairman of the Executive Council, Weisgal asserted that, despite these voices, the decision was adopted to follow Weizmann's vision and to go on with the expansion of the institute. He thus wrote:

> Purely political work for the establishment of a Jewish national homeland, without a solid basis of practical building and economic development in Palestine, would have been valueless… The most vital contribution, which World Jewry could make towards a new society in Palestine would evolve through the enrichment of cultural elements and the creation of a sound basis of scientific advance … Weizmann … did not interpret culture and science as an abstract philosophy of life. He and those who followed his leadership envisaged the development of the Zionist up building effort in terms of concrete value … To us, the sponsors and friends of the project, the Weizmann concept represented the best guarantee of the eventual realization of the true purpose of Zionism.[21]

And indeed, plans for expansion continued to be carried out. The cornerstone laying ceremony for the Institute of Physics and Physical Chemistry took place in July 1946. By that time, the construction plans as well as the scientific program for

---

[20]Bergmann, "The achievements of Israeli science," *DAVAR Apr. 4*, 1958, 23. (Hebrew).

[21]Weisgal, "Report by the Chairman of the Executive Council for the Period from November 2, 1949 to Jun. 30, 1952 (draft)"; (WIA 12–90–17).

Fig. 2.1 Chaim Weizmann with his secretary Meir Weissgal (standing behind him) dedicate the embossed cornerstone of the Weizmann Institute in Rehovot (June 6, 1946). Courtesy of the Israel National Photo Collection (*Credit* Hans Pinn)

the new institute were already fairly detailed (Fig. 2.1). The actual construction activity progressed in parallel with the dramatic political events in Palestine, which, following the United Nations Partition Plan for Palestine, culminated in a military conflict between Israel and the Arab states.

In Weizmann's Zionist doctrine, as already indicated, political activity aimed at creating a Jewish national home would be worthless without a solid foundation of practical construction and economic development in Palestine (Figs. 2.2 and 2.3). The establishment of the state strengthened this view, because, together with the massive immigration, the problems of economic survival became increasingly pressing. Israel was now facing new challenges in struggling for its economic independence, challenges that science could help address with great success. Moreover, since the founding of the Sieff Institute, the connection between this institution and a practical Zionist program had been a guiding principle for Weizmann. The institute, as Weisgal was fond to emphasize, was for Weizmann "another 'political fact' and, even more than that, a tool to help in creating the scientific foundations of the national structure."[22] Weisgal added that for Weizmann, the importance of establishing the Institute lay in the development of science in a deprived environment, in the creation of equipment and the work environment, in the creation of a scientific atmosphere and in the establishment of a

---

[22]WIS, "Scientific Activity Report 1953," (WIA).

Fig. 2.2 Weizmann, with Arieh Elhanani and Israel Dicker, the Institute architects, in the building site (1947). Courtesy of Nima Geffen

Fig. 2.3 Weizmann, with Arieh Elhanani and Israel Dicker, the Institute architects, in the building site (1947). Courtesy of Nima Geffen

scientific tradition. Preserving a research environment in a semi-wild place, as well as maintaining contact with developments in the distant world of science, were the main goals of the research institute as now formulated by Weisgal.[23]

When the war broke out in 1948, the facilities of the Sieff Institute and those of the partially completed Weizmann Institute were made available to the army. The war effort was supported by scientists from Rehovot, the Technion in Haifa and the Hebrew University. Weisgal claimed that this involvement was initiated in full agreement with the American Committee, as well as with Weizmann himself. He also declared proudly, some years later, that WIS played a decisive role in the outcome of the war:

> A scientific institution, in the United States for example, emerged richer rather than poorer from a war. When the Israel war ended last year, we emerged much poorer; in fact our million dollar reserve had been eaten up by the war.[24]

But in fact, Weizmann was far from satisfied with the military involvement of WIS. He expressed his dissatisfaction explicitly in a letter to Weisgal, from July 1948, in the following terms:

> There is no reason why the whole of the scientific work, into which so much energy, devotion, and love has been poured, should be replaced by something which is not science but making explosives.[25]

On November 2, 1949, exactly five years after taking the decision to expand the Sieff Institute, the inauguration ceremony of WIS was held. The new institution in Rehovot was thus joining two centers of higher education and advanced research that had already been established in Mandatory Palestine: the Hebrew University in Jerusalem and the Technion in Haifa, both of which opened their gates to students in 1925. The Board of Directors of WIS comprised its administrative director, the physicist Benjamin M. Bloch (1900–1959), its scientific director, Bergmann, and the chairman, Weisgal. Weizmann, who was the president of the Sieff Institute since its founding, became the president of the Weizmann Institute of Science,[26] as well as the first president of the State of Israel (Samuel 1970) (Fig. 2.4).

In his speech at the official ceremony, Weizmann adopted the rhetoric of the most fervent Zionist ethos of the time to express his views about science:

> We live, as you know, in a pioneering country. We are pioneering in the wilderness, in agriculture, and in industry. But here in Rehovot we are also engaged in a peculiar kind of pioneer work—we are pioneering in science. There are many problems to be solved in our land, and many difficulties to be overcome. There are also many dangers still to be met. But

---

[23]Weisgal, "Report by the Chairman of the Executive Council for the Period from November 2, 1949 to Jun. 30, 1952 (draft)" (WIA 12–90–17).

[24]Weisgal, "Report by the Chairman of the Executive Council for the Period from November 2, 1949 to Jun. 30, 1952" (draft) (WIA 12–90–17).

[25]Weizmann to Weisgal, Jul. 30, 1948 (LPCW, Series A, Vol. 22, letter 238).

[26]Kaufman, Edmund I.; Stone, D. Dewey; Levine, Harry; Weisgal, Meyer, Second annual report (WIA 24–76–6).

Fig. 2.4 Chaim Weizmann and his closests associates: Bergmann, Weisgal and Bloch (Feb. 18, 1949). Courtesy of the Israel National Photo Collection

to meet them, we must not rely only or chiefly on physical force. We have a mighty weapon which we must utilize with ingenuity and skill, and with every means available to us. Science is that weapon, our vessel of strength and our source of defence.[27]

The issue of basic-theoretical versus applied science continued to surface in various contexts, and in the first scientific activity report of WIS, Weisgal expressed the mixed emphasis of its leaders, on this topic:

The reader will note throughout the natural fusion between pure science and applied science; he will also note that there are practical scientific tasks which have a worldwide application but are of special import to the State of Israel.[28]

But a few years later, in the 1955 report, Weisgal returned to this issue with a different emphasis that indicated the directions into which WIS choices were now more consistently driven:

Take, as an example, the stormy debates on pure versus applied science that rocked the walls of our Institute up to about four years ago. All of this conflict has evaporated into the void, disposed of by the exigencies of ever-evolving life.

Let me confess that I have never been too clear as to the distinction between "pure" and "impure" science. I was inclined to accept the primitive classification that science was

---

[27]Weizmann, "Speech in the Dedication Ceremony of WIS," Nov. 2, 1949. Quoted in Rose (2015, Kindle Locations 9259–9264).

[28]WIS, "Scientific Activity Report 1949" (WIA).

"pure" as long as it remained in the realm of theory. Since then I have learned that the pure research of today is only the applied science of tomorrow.[29]

Weisgal was of the view that WIS should enlist in the service of the state in order to overcome the national economic crisis. In this sense, his views were somewhat similar to those expressed by Bergmann, though in a more moderate fashion. He tried to find ways to minimize the negative impact of any involvement with security matters on more purely scientific activities that would lead to a prompt recognition of the Institute as a prestigious scientific institution in the world. While he admitted that technological development was of the greatest importance for transforming the State of Israel into a strong independent entity, he did not adopt the more extreme, security-oriented position of which Bergmann became so strongly partisan. In times of emergency, he favored a more intense involvement, but right after the end of the war he returned to the view that the Institute should focus in the first place in achieving the highest standards of scientific accomplishment and this would be the best way to serve the newly-born state, its economy and its society. In addition, Weisgal also maintained his full personal allegiance to Weizmann, at a time when Bergmann became increasingly close to Ben-Gurion. In terms of a project like the WEIZAC, a technological engineering project with the potential of contributing to the advancement of science as well as to the economic development of the country, Weisgal's approach played a very favorable role. Given his central role in the decision-making processes at WIS and his personal proximity to Weizmann, Weisgal proved to be a key player in the success story of WEIZAC.

The Weizmann Institute was established in Rehovot, so it happened, right at the time when the IAS machine was being built at Princeton. The crucial turning point in the story of WEIZAC, at the confluence of these two separate threads, came when Chaim Leib Pekeris, joined in 1946 the Scientific Planning Committee in the USA. Pekeris immigrated in 1948 and brought with him to Rehovot the idea of building an automatic computing machine. He also brought an unusual capacity for implementing the daring project. The next section is devoted to describing Pekeris's scientific background and personality, and the initial stages of the process that led to the unlikely decision to build an electronic computer in Rehovot during the early years of the State of Israel.

2.4 Pekeris, Applied Mathematics, Mathematical Brains

Chaim Leib Pekeris (Fig. 2.5) was born in Lithuania and already at an early age he stood out for his intellectual skills. His uncle, who had settled in the United States, helped him and his two younger brothers to immigrate and to continue their studies there. One of Pekeris's sisters immigrated to Palestine in 1935. His parents and another sister remained in Europe and they were all killed in the Holocaust. In

---

[29]WIS, "Scientific Activity Report 1955" (WIA).

Fig. 2.5 Portrait of Chaim L. Pekeris, Dean of the Faculty of Mathematics at the Weizmann Institute in Rehovot (Jan. 1, 1973). Courtesy of the Israel National Photo Collection (*Credit* Fritz Cohen)

1925, Pekeris received a degree in mathematics from MIT. He continued with graduate studies in the department of aeronautical engineering, specializing in meteorology, which was then a new discipline. He completed his master's degree in 1929 and his doctorate (Sc.D.) in 1933. His advisor was Gustav Rossby (1898–1957), the de facto director of the meteorological project in Princeton and one of the world's leading meteorologists at the time. Working in the department of geophysics at MIT until 1940, Pekeris established his position as a promising young scientist. He made creative contributions to geophysics, astrophysics and hydrodynamics (Freeman 2004). This were all rather new fields of enquiry in applied mathematics, in which intensive computations played a central role, as will be discussed below.

During World War II, Pekeris was involved in military research as a member of Columbia University's Hudson Laboratories and he investigated the propagation of acoustic waves and pulses. For his wartime research, Pekeris was recognized by the U.S. Navy with the title of "honorary admiral" (Freeman 2004, 221). The assignment at Columbia allowed him to work, hands-on, with a state-of-the-art electronic computer, the Relay Interpolator—later named Model II—that had been operational at Bell Labs in New York since September of 1943 (Irvine 2001). Pekeris described retrospectively in 1987 his first impression from the automatic computer in the following words:

> In the middle of the war there was a team of the Division of War Research at Columbia University located on 64th floor of the Empire State Building. One day I was told that we had available a device that does computations in Bell Telephone Research Laboratories, in downtown New York. I went down there and I saw a computer built on relays, mechanical relays ... The thing that impressed me most at first is how the most complicated mathematical operations can be built on these simple elements, yes or no. To this day it's a remarkable thing.[30]

After the war, Pekeris became head of the mathematical physics group at Columbia University. During these years, he was involved in additional research of seismic wave propagation. After the Office of Naval Research had established the meteorological project under the direction of von Neumann at the IAS in Princeton, Pekeris was invited in 1946 to participate as a consultant.[31] The initial aim of the project was to examine the potential use of the electronic computer in theoretical meteorology research and in weather forecasting. Quite remarkably, from the proposal of the project one gets the impression that those involved intended not just to be able to forecast but, with the help of the computer, even to take the first steps towards influencing it "by rational, human intervention."[32]

It is fair to assume that participation in this project deepened Pekeris's understanding of the digital computer's capabilities, and strengthened his personal relationship[33] with von Neumann (Harper 2008, 98–121). Combined with his earlier involvement with intensive-calculation research in various fields of applied mathematics, Pekeris developed a rather peculiar kind of scientific profile, which few people at the time could boast. Thus, it was far from obvious for a successful

---

[30]Lee Segel, Conversation with Pekeris, Feb. 24, 1987 (HMF).

[31]Pekeris to Ettlinger (University of Texas), Dec. 13, 1946 (CPA).

[32]Frank Aydelotte (the IAS director) to LCDR, Daniel F. Rex of the Office of Naval Research, 8.5.1946 Quoted in Harper (2008, 103).

[33]Rubinoff reported that "Pekeris in fact was a rather close friend of Johnny von Neumann's" (Rubinoff, Interview by Richard R. Mertz, May 17, 1971, (SOVA,1969–1973, 1977). (https://sova.si.edu/record/SIA.FA06-010?s=0&n=10&t=C&q=oral+history&i=0) A year before his death, von Neumann wrote to Pekeris: "I am really touched by the signs of true friendship that you are giving me". (von Neumann to Pekeris, Feb 24 1956 (CPA)).

scientist of his kind to consider the possibility of abandoning the leading institutions that afforded him the few natural settings in existence at that time, where he could continue to develop fruitfully his skills and talents, in order to move to an adventurous, and by all means peripheral environment, in the newly created Weizmann Institute at Rehovot. Clearly, a main motivation for taking this step is to be found in his sincere commitment to the Zionist ideology. Indeed Pekeris was a declared Zionist "all his life."[34]

Pekeris was not alone in making this kind of bold move at the time. A considerable part of the budding scientific activity in Mandatory Palestine was set up by Jews who immigrated in the first half of the twentieth century. They played a significant role in constructing the scientific core of the Hebrew University, the Technion, and the Sieff Institute. Some of them came to Palestine following their Zionist ideology and some did so for lack of a better choice, as they escaped from Europe. In 1923, for instance, Professor Andor Fodor (1884–1964) came from Germany to establish the Institute for Chemical Research at the Hebrew University (Deichmann and Travis 2004). The brilliant German mathematician Edmund Landau (1877–1938) arrived from the University of Göttingen and helped to establish the Einstein Institute of Mathematics also in Jerusalem. He left after about eighteen months due to different personal and professional reasons, and he was replaced by Abraham Halevi Fraenkel (1891–1965) who came from Kiel, Germany, and by Michael Fekete of Budapest (1886–1957) (Katz 2004). As already mentioned, also Bergmann immigrated to Palestine in the 1930s. Another example worthy of mention is that of Frantz Ollendorff (1900–1981), who arrived from Germany to join the Technion in 1937, and in 1938 he established the Faculty of Electrical Engineering there (Baal-Schem 2007).

As early as 1936 Pekeris had paid a visit to Palestine. At that time he unsuccessfully looked for an appointment in some of the existing institutions. After a few years, in October 1945, Pekeris applied for a position in the Hebrew University of Jerusalem and sent his resume to Leon Roth (1896–1963), who was Professor of Philosophy at the Hebrew University and at the University of Manchester, and later Rector of the Hebrew University during 1940–1943.[35] Two prominent scientists supported Pekeris's application. One of them was the physical-chemist Adalbert Farkas (1906–1995), active in Jerusalem since 1936. Farkas wrote to the administrator of the Hebrew University, David Werner Senator (1896–1953), informing about the desire of his old friend, Pekeris, to move to Palestine provided he could get an academic position. Farkas was clear in expressing his recommendation:

> I don't think there is any need to point out the splendid opportunity for the university to secure such a valuable man as Prof. Pekeris,…His ability would be valuable to the country as a whole as well as the university…[36]

[34]Lee Segel, Conversation with Pekeris, Feb. 24, 1987 (HMF).

[35]Pekeris to Leon Roth, Mar. 13, 1945 (WIA 3–96–98).

[36]Farkas Adalbert to David Werner Senator, Sep. 20, 1944 (HUA—165 Pekeris).

The other was the Caltech mathematical physicist Paul S. Epstein (1883–1966), who recommended Pekeris "most warmly."[37] Still, the application was rejected.[38] Bergmann, who had met Pekeris in New York, foresaw this rejection and thus wrote to Weizmann with the intention of bringing Pekeris to the new institute.[39] Pekeris reacted enthusiastically to the offer and indeed joined the founding team of WIS in 1946.[40] This turned out the first practical step on the way to the construction of an early powerful electronic computer in the new State of Israel.

In his application letter, Pekeris emphasized the importance of the discipline of "applied mathematics" in which much progress had been made during World War II. Pekeris listed some newly established institutes and centers for applied mathematics around the world, based on massive calculations, such as the Institute of Applied Mathematics at Brown University, the Applied Mathematics Center at MIT, Harvard, and Columbia. More specifically, he wrote about the need for automated computation and reported about:

> … the installation in several Eastern Universities and in industrial laboratories as well as government research divisions, of computing machines, popularly known as mathematical brains. (The latest such "brain" now under construction[41] will cost more than $200,000).[42]

We already mentioned above the issue of pure, or basic, versus applied science in relation with the question of the desired overall orientation for WIS. The term "applied mathematics" as intended by Pekeris and by his contemporaries in this context, involves a more specific matter that requires some clarification. The relationship between the two opposed but at the same time complementing aspects of the discipline, pure and applied mathematics, has sometimes been described as one of human attitudes and motivation, "pure mathematics is directed towards logical crystallization, abstraction, generalization; applied mathematics means close interconnection of mathematical methods with physical reality" (Courant 1956, 1). Though somewhat simplistic, this is a useful characterization. Of course, mathematical ideas have been applied in practical contexts all along history, but a more substantial idea of "applied mathematics" developed from the seventeenth century on, as the new infinitesimal calculus, on the one hand, and Newtonian physics, on the other hand, became central paradigms of science. Disciplines like analytic mechanics, kinetic theory of gases, hydrodynamics, and

---

[37]Paul S. Epstein to Leon Roth, Apr. 20, 1945 (HUA—165 Pekeris).

[38]Michael Fekete to Pekeris, Oct. 10, 1945 (HUA—165 Pekeris).

[39]Bergmann to Weizmann, Jun. 22, 1945 1945 (CWA—27–2590).

[40]Kaufman, Edmund I.; Stone, D. Dewey; Levine, Harry; Weisgal, Meyer, Second annual report (WIA 24–76–6); Weisgal, Report to the Board of Directors of the American Committee and to Committees in Other Countries for the Weizmann Institute for the Period of 1944–1949, 10 (WIA). In this report, Pekeris is formally described as head of the DAM.

[41]Pekeris knew about Mark I (ASCC) in Harvard as well as RCA and Bell Labs efforts.

[42]Pekeris to Leon Roth, Mar. 13, 1945 (WIA 3–96–98).

electrodynamics flourished under this classic paradigm. Roughly speaking, towards the end of the nineteenth century, France and Britain dominated the scene of applied mathematics in this classical sense, whereas Germany focused more strongly on the pure branches of mathematics such as analysis and number theory. To the extent that it is valid to speak about this distinction, and of its national dimensions, it corresponded to well-established traditions of research as well as to ideological concerns. Thus, the neo-humanism preached by Wilhelm von Humboldt (1767–1835) and on which he established the Berlin University at the turn of the nineteenth century, provided a consistent framework for the kind of purist approach followed in that institution, as well as in various others in the German speaking countries (Pyenson 1983).

Between 1890 and the end of World War II, new mathematical tools were developed, mainly in the field of mathematical modelling, stochastic processes and statistics, which gave rise to new kinds of applications, in fields such as electrical communications, aviation, economics, biology, political science, and psychology. The rise of the electronic computer after World War II infused further impetus upon these trends, mainly in disciplines based on intensive computations such as meteorology, geophysics, oceanography and others (Barrow-Green and Siegmund-Schultze 2015). The latter were the fields in which Pekeris and other mathematicians of similar inclinations excelled and started to develop their own careers.

It is important to notice, however, that the adoption of automatic computers as an efficient new tool for various branches of science, and as an important way for implementing numerical methods in various scientific disciplines, or in fields of pure mathematics, was far from being self-evident or straightforward (Corry 2008, 40–48). Illustrative evidence of the challenges that arose as part of this interesting process is found in a text by Leslie John Comrie (1893–1950), a pioneer of scientific computation using commercial calculators and other tools of computation. Comrie headed several important scientific projects, such as the Computing Section of the British Astronomical Association, between 1920 and 1922 (Croarken 2003). These were computation-intensive projects that involved the coordinated efforts of large teams of human calculators who were assigned specific, narrowly defined tasks, and who performed them individually, either manually or with the help of somewhat rudimentary, mechanical desktop calculators. This was the high point of a period of time "when—as Alan Grier aptly defined it—computers were still human" (Grier 2013; See also Corry 2017). Only a few *automatic computers* existed, and they were different from today's high-speed computers, as the stored-program concept had not yet matured at that time.

Comrie, always in search of innovative, improved methods of calculation, admired the new electronic computers when they started to appear, but he thought that their high prices, as well as the difficulty of programming them, did not justify their construction and their use in most scientific calculation problems. In his opinion, the computer was a futuristic technology that was not yet ripe. In 1946 he wrote an interesting review on this topic in the journal *Mathematical Tables and*

Other Aids to Computation. This journal was the main international venue of publication for anyone interested in the kind of scientific computational projects in which Comrie and his colleagues were involved. The creation and early stages of the journal embodies in itself the important transformations underwent by the attitudes of the mathematical community concerning the role of electronics computers in their discipline (Corry 2010). In reviewing the use of commercial computing machines for scientific purposes, Comrie wrote:

> I am convinced that the day of the desk machines is not yet over or even threatened by the half dozen or so large and special machines that have come into being during the war. Nevertheless, I join with others in admiring these machines, and, after seeing so much binary multiplication, feel that LEWIS CARROL should be alive now to write Alice in Onederland. There is however much more to be done before the usefulness to science of the commercial machine is exhausted (Comrie 1946).

Pekeris's efforts led to the adoption of the ethos of "applied mathematics" as the leading one at WIS. This is especially remarkable when one considers the fact that, from its inception, the Einstein Institute of Mathematics at the Hebrew University advocated the pursuit of the most stringent purist tradition along the lines of the Berlin school. The founding fathers, Landau, Fraenkel and Fekete, saw themselves as "proud intellectual inheritors of this variety of the Berlin tradition that not only conceived pure mathematics as a sublime neo-humanistic ideal, but in parallel also disdained applied mathematics" (Katz 2004). But, in addition, the rhetoric that accompanied this scientific ideology also emphasized its adequacy to the kind of "spiritualized" brand of Zionism that they saw themselves as representatives of, and of which the creation of the Hebrew University was the high point (Corry 2010).

For Landau himself, the creation of a chair or mathematics in Jerusalem embodied a natural combination of the Jewish traditional ethos of studying Torah for its own sake, in which his own ancestry excelled, with the modern ethos of science for its own sake, of which he was a preeminent figure. When he assumed the chair in 1927, a Yiddish newspaper in New York described the situation in the following, highly emotive terms:

> The wheel of history comes around. About one hundred and fifty years ago, there lived in Prague the Ga'on Rabbi Yehezqel Landau of blessed memory, called by scholars "Noda' bi-Yehudah." The son of "Noda' bi-Yehudah," Rabbi Shmu'el Landau, was well known as a great scholar. The grandson of "Noda' bi-Yehudah," Rabbi Mosheh Landau, wrote the book Ma'arakhey Lashon. Today all their descendants are either completely or partly assimilated. However, in this time of (our) salvation, another genius has emerged among his descendants. But this time—a modern genius. One of the four greatest mathematical geniuses of the whole world. And he is a professor from the extolled Göttingen University, Dr. Yehezqel Landau, who bears the name of his great-grandfather. In the forthcoming winter semester he will be lecturing his lessons in high mathematics before students at the Hebrew University in Jerusalem. He has been drawn back to his original source, as if his great-grandfather gripped him by the nape of his neck and brought him to lecture here in Jerusalem. As he himself said, he feels that "the Torah has gone forth from Zion." Professor Landau continues in Jerusalem the chain begun by his great-grandfather, Rabbi Landau of Prague (Cited in Katz 2004, 213).

Applied mathematics was not on the agenda of the Hebrew University for decades to come, in spite of the fact that, right from the beginning, it was proposed and seriously considered. In the opening ceremony of the building for the Mathematical Institute, on April 1925, the British applied mathematician Selig Brodestky (1888–1954) (who later on, in 1949, served for a very short period as president of the university) advocated for the importance of research in applied mathematics with the following words:

> The applied mathematician fulfills the important task of acting as the bond of union between physics and mathematics. The applied mathematician brings to the aid of the physicist the mathematical equipment to solve the problems encountered by the physicist. He also brings to the [pure] mathematician knowledge of realities of nature and prevents him from losing himself in barren speculations (Cited in Katz 2004, 220).

By the late 1940s, this kind of rhetoric was still foreign to the main concerns of the Jerusalem mathematicians but not so to Pekeris, who perceived applied, computation-intensive mathematics as a substantial branch of legitimate, cutting-edge research, and in particular the kind of research that justified the expenses of building of an electronic computer in Rehovot. For him, it was a matter of scientific convictions, that this, and not the purist approach typical of the Jerusalem school, was the kind of research worthy of pursuit at WIS.[43] But at the same time, this was the kind of research that, in the views of Pekeris, his colleagues, and the leaders of WIS, was the more suitable one to the aims of Zionist as conceived by them. In the transitional period between the Mandate and the first years of the state, this became an even more strongly preached view. By diverging from the purist ideology of the Jerusalem school, the emphasis on the applied, computation-intensive style of Pekeris helped turning WIS into a center of scientific excellence also in mathematics from very early on, and in its own way.

Already in his first years at WIS, Pekeris came up with an important contribution along his style of applied mathematics as he became directly involved with the geological survey of the Land of Israel.[44] He had a firm opinion about the importance of engineering for the newly born state, which he expressed publicly and consistently in every possible occasion. Thus, for instance, in 1966, while addressing the second conference of IPA, Information Processing Association of Israel, he said:

> … Professor Theodor von Karman[45] … said… they—the gentiles—say that we Jews are not qualified in the engineering profession. And there were days when we heard the same

---

[43]Estrin, Interviewed by Mapstone, Robina, 15 Jun. 1973. (Smithsonian, Computer Oral History Collection #96, Box 6, Folder 12).

[44]WIS, "Scientific Activity Report 1953," 37–38 (WIA).

[45]Theodor von Karman (1881–1963) was a Hungarian-born, German-educated, American research engineer, best known for his pioneering work in aeronautics. During the 1940s and the 1950s he was involved in establishing the department of aeronautics engineering department at the Technion, Haifa. Singer, J. History, Faculty of Aerospace Engineering. 2006. http://aerospace.technion.ac.il/department/history (accessed Feb. 18, 2018).

argument regarding agriculture and military service, and the Zionist movement came and changed it… We cannot ignore the fact that today we have another Zionist mission, to instill engineering in the Jewish intelligentsia…

We do not succeed in engineering because we do not value it… The Jews do not succeed in engineering because they do not hold this profession. I believe that these days the problem is essential, essential to the very existence of the state, and perhaps to the whole nation. We have to do applied work in engineering (Pekeris 1966, 12).

In a meeting of the Scientific Advisory Committee, which dealt with future programs for WIS, Pekeris talked about the technological revolution that was taking place in the country, and urged WIS to take a lead in this important process. He agreed with the chairman of the meeting, Weisgal, who said that:

Unless the Institute now becomes a participant in this revolution, it would be left behind to become an "ivory tower," with no relationship whatsoever to what was happening in Israel.[46]

In this meeting, Pekeris proposed to increase the number of engineers in the WIS staff, and even create some kind of engineering school.

On November 1947, Pekeris was formally appointed head of the Department of Applied Mathematics (DAM) of WIS.[47] However, his arrival was delayed due to his involvement in purchasing war equipment for the "Yishuv" in Palestine.[48] After the establishment of the State of Israel, at the end of 1948, Pekeris settled in Rehovot and took the position in which he became the major leading figure of the WEIZAC project.

References

Achuthan S, Agrawal BC, Vimal SP, Thakore SR (1992) Computer technology for higher education: the Indian experience, vol II. Ashok Kumar Mittal, New Delhi

Arora A, Gambardella A (2006) From underdogs to tigers: the rise and growth of the software industry in Brazil, China, India, Ireland, and Israel. Oxford University Press, Oxford

Aspray W (1985) New introduction by William Aspray. In: Proceedings of a symposium on large-scale digital calculating machinery by the Harvard Computation Laboratory (1947), vol ix. MIT and Tomash Publisher, Cambridge

Aspray W (1986) International diffusion of computer technology, 1945–1955. IEEE Ann Hist Comput 8(4):351–360

Aspray W (1990) John von Neumann and the origins of modern computing. MIT Press, Cambridge

---

[46]SCM held on Aug. 12, 1968 (CPA).

[47]Weisgal to Pekeris, Nov. 13, 1947 (WIA 6–74–24). A department of "pure mathematics" was established at the Institute only in 1969.

[48]"Annual Meeting of the American Committee of Weizmann Institute of Science," Dec. 8, 1948 (WIA 6–74–24); D. Ben-Gurion, Diary, Sep. 9, 1948, (BGA); (Freeman 2004).

Baal-Schem J (2007) The birth of a hi-tech society: first steps in electronics and computing in Israel. In: EUROCOM 2007. The International Conference on "Computer as a Tool". IEEE, pp 2638–2640

Ball BL, Kessel L, Pearl C (1996) The encyclopedia of Jewish life and thought. Digitalia, Jerusalem. eBook Collection (EBSCOhost), EBSCOhost

Banerjee, UK (1996) Computer Education in India: Past, Present and Future. Concept Publishing Company

Barrow-Green J, Siegmund-Schultze R (2015) The history of applied mathematics. In: Higham NJ (ed) The Princeton companion to applied mathematics. Princeton University Press, Princeton, pp 55–79

Berenblum I (1966) Creating a scientific tradition in a new country. In: Victor E (ed) Meyer Weisgal at seventy – an antology. Weidenfeld and Nicolson, London

Bigelow J (1980) Computer development at the Institute for Advanced Study Princeton. In: Metropolis N, Howlett J, Rota G-C (eds) A history of computing in the twentieth century. Academic Press Inc., London, pp 291–310

Blachman MN (1953) A survey of automatic digital computers. Office of Naval Research, Washington D.C.

Breznitz D (2007) Innovation and the state. Yale University Press, New Haven

Burks AW, Goldstine HH, von Neumann J (1946) Preliminary discussion of the logical design of an electronic computing instrument. https://library.ias.edu/files/Prelim_Disc_Logical_Design.pdf. Accessed 1st Jan 2019

Calder R (1962) The secret of life. In: Weisgal MW, Carmichael J (eds) Chaim Weizmann: a biography by several hands. Weidenfeld and Nicolson, London, pp 114–125

Campbell-Kelly M, Aspray W (eds) (2004) Computer: a history of information machine. Westview Press, Boulder

Ceruzzi PE (2003) A history of modern computing. MIT Press, Cambridge

Cohen U (2016) From political rejection to scientific renewal: Weizmann and the establishment of Daniel Sieff research institute in Rehovot, 1931–1934. In: Cohen U, Chazan M (eds) Weizmann The leader of Zionism. Zalman Shazar Center for Jewish History[Hebrew], Jerusalem, pp 380–383

Cohen U, Chazan M (2016) Weizmann the leader of Zionism. Zalman Shazar Center for Jewish History [Hebrew], Jerusalem

Comrie LJ (1946) The application of commercial calculating machines to scientific computing. Math Tables Other Aids Comput 2(16):149–159

Corry L (2008). Fermat meets SWAC: Vandiver, the Lehmers, computers, and number theory. IEEE Ann Hist Comput 30(1):38–49

Corry L (2010) Hunting prime numbers, from human to electronic computers. Rutherford J N Z J Hist Philos Sci Technol 3

Corry L (2017). Turing's pre-war analog computers: the fatherhood of the modern computer revisited. Commun ACM 60(8):50–58

Corry L, Schappacher N (2010) Zionist internationalism through number theory: Edmund Landau at the opening of the Hebrew University in 1925. Sci Context 23(4):427–471

Cortada JW (2013) How new technologies spread: lessons from computing technologies. Technol Cult 5(12):229–261

Courant R (1956) Methods of applied mathematics. In: Shamos MH, Murphy GM (eds) Recent advances in science, physics and applied mathematics. Science Editions, New York, pp 1–14

Croarken M (2003) Table making by committee: British table makers 1371–1965. In: Campbell-Kelly R, Croarken M, Robson E (eds) The history of mathematical tables: from Sumer to spreadsheets. Oxford University Press, Oxford, pp 235–267

Deichmann U, Travis AS (2004) A German influence on science in mandate Palestine and Israel: chemistry and biochemistry. Isr Stud 9(2):34–70

Dyson G (2012) Turing's cathedral. The origin of the digital computer. Pantheon Books, New York

Estrin G (1952) A description of the electronic computer at the institute for advanced studies. In: Proceedings of the 1952 ACM national meeting (Toronto). ACM, New York, pp 95–109

Fischer L (1994) Chaim Weizmann: the first president's selected documents. Government Printer [Hebrew], Jerusalem

Freeman G (2004) Biographical memoires V.85. National Academies Press. http://www.nap.edu/catalog/11172.html. Accessed 2 Feb 2010

Godfrey MD, Hendry DF (1993) The computer as von Neumann planned it. IEEE Ann Hist Comput 15(1):11–21

Golani M, Reinharz J (2019) The founding father. Chaim Weizmann. A biography (1922–1952). Magnes Press [Hebrew; Forthcoming.], Jerusalem

Goren AA (1997) Juda L. Magnes and the early years of the university. In: Katz S, Heyd M (eds) The history of the Hebrew University in Jerusalem. Origins and beginings. The Hebrew University Magness Press, Jerusalem, pp 363–387

Grier DA (2013) When computers were human. Princeton University Press, Princeton

Haigh T (2016) ENIAC in action: making and remaking the modern computer, Kindle edn. History of computing. MIT Press, Cambridge

Haigh T, Priestley M, Rope C (2014) Reconsidering the stored-program concept. IEEE Ann Hist Comput 36(1):4–17

Harper KC (2008) Weather by the numbers. The genesis of modern meteorology. MIT Press, Cambridge

Hartree D, Newman M, Wilkes M, Williams F, Wilkinson J, Booth A (1948) A discussion on computing machines. Proc R Soc Lond Ser A Math Phys Sci 195(1042):265–287

Irvine MM (2001) Early digital computers at Bell Telephone Laboratories. IEEE Ann Hist Comput 23(3):22–42

Jensen WB, Fenichel H, Orchin M (2011) Scientist in the service of Israel: the life and times of Ernst David Bergmann (1903–1975). Hebrew University Magnes Press

Katz S (2004) Berlin Roots-Zionist incarnation: the ethos of pure mathematics and the beginnings of the Einstein Institute of Mathematics at the Hebrew University of Jerusalem. Sci Context 17 (1–2):199–234

Katz S, Heyd M (1997) History of the Hebrew University of Jerusalem, origins and beginings. The Hebrew University Magnes Press, Jerusalem

Keidar A (1976) Brit Shalom: the early period (1925–1928). In: Bauer Y et al (eds) Studies in the history of Zionism. Hasifirya Hatzionit, [Hebrew], Jerusalem, pp 224–283

Kirsh N, Katzir S (2016) Between chemistry and politics: Weizmann's scientific activity during the 1930's and 1940's. In: Cohen U, Chazan M (eds) Weizmann the leader of Zionism. Zalman Shazar Center for Jewish History [Hebrew], Jerusalem, pp 413–440

Kolatt I (1997) The idea of the Hebrew University in the Jewish national movement. In: Katz S, Heyd M (eds) The history of the Hebrew University of Jerusalem. Hebrew University Magness Press, Jerusalem, pp 3–74

Kotzin DP (2010) Judah L. Magnes, An American Jewish nonconformist. Syracuse University Press, Syracuse

Mahoney MS, Haigh T (2011) Histories of computing. Harvard University Press, Cambridge, MA

Parzen H (1970) The Magnes-Weizmann-Einstein controversy. Jewish Soc Stud 32(3):187–213

Pekeris CL (1966) Greetings. In: Proceedings of the National Conference on Data Processing. IPA —Information Processing Association of Israel, Jerusalem, pp 12–13

Priestley M (2011) A science of operations: machines, logic and the invention of programming. Springer, New York

Prokhorov SP (1999) Computers in Russia: science, education, and industry. IEEE Ann Hist Comput 21(3):4–15

Pyenson L (1983) Neohumanism and the persistence of pure mathematics in Wilhelmian Germany. American Philosophical Society

Rao P (2008) TIFRAC, India's first computer—a retrospective. Resonance 13(5):420–429

Reinharz J (1993) Chaim Weizmann: the making of a statesman, vol 2. Oxford University Press, New York

Reinharz J (1997) The speech of Dr. Weizmann: commentary. In: The history of the Hebrew University: origins and beginings. Hebrew University Magness Press, Jerusalem, pp 323–326

Rose N (2015 [1987]) Chaim Weizmann: a biography, Kindle edn. Endeavour Press. (First edn. Weidenfeld and Nicolson, 1987.)

Rosenkranz Z (2011) Einstein before Israel: Zionist icon or iconoclast?. Princeton University Press, Princeton

Samuel R (1970) A Brief history of the institute. In: Shultz L (ed) Gateway to science: the Weizmann Institute at twenty-five. Weizmann Institute, Rehovot

Sean O (1997) The birth of a Celtic tiger. Commun ACM 40(3):11–16

Sieff I (1970) The memories of Israel Sieff. Weidenfeld and Nicolson, London

Tidhar D (1959) Encyclopedia of the founders and builders of Israel, vol 10. [Hebrew]

Tinn H (2010) Cold war politics: Taiwanese computing in the 1950s and 1960s. IEEE Ann Hist Comput 32(1):92

von Neumann, J (1945) First draft of a report on the EDVAC. University of Pennsylvania, Moore School of electrical Engineering. Reprinted in IEEE Ann Hist Comput 15(1):27–65 (1993)

Wagner S (2015) The Zionist movement in search of grand strategy. J Mil Strat Stud 16(1):61–89

Weisgal MW (1971) So far: an autobiography. Transaction Publishers, New York

Weizmann C (1936) The connection between Thora and action, speech at the party of the friends of the Hebrew University at Tel Aviv, 19.2 1936. In: Chaim Weizmann on the Hebrew University. Haivry [Hebrew], Jerusalem, pp 36–37

Weizmann C (2013) Trial and error: the autobiography of Chaim Weizmann, vol 2, Kindle edn, (1918–1948). Plunkett Lake Press

Chapter 3
The WEIZAC Challenge: Building an Electronic Brain in Rehovot

Pekeris's plans for the construction of a computer as the centerpiece of the activity of his department required, above all, gaining the active support of the key figures at the Institute, and in the first place Weizmann, Bergmann and Weisgal. Pekeris intelligently built their faith in his ability to carry out successfully the project in the highly unfavorable conditions of the newly created state. In particular, through Weisgal's support, he was able to ensure the required funds. During the planning years of WIS, with Pekeris still in the USA, he directed his efforts at reaching the initial approval for the project. Later on, after his arrival at WIS, his actions were carefully planned to set the stage and indeed to create a "sense of inevitability"[1] for the electronic computer in Israel. His strategy was successful by any standard. During the first years of the Institute (1949–1951), despite the adverse material conditions, Pekeris continued to push the project forward until the decision was finally reached and the computer became a reality. In the background, Pekeris worked hard to shape the conditions required to initiate and successfully complete the project.

Since the establishment of WIS, Pekeris mentioned the computer project everywhere and in every opportunity: in discussions about manpower, in work plans of the DAM, and even in future plans of other WIS departments. When the computer project was finally discussed at the meeting of the Scientific Committee in January 1952, it was easily approved, without any thorough discussion and almost without objection.[2]

---

[1]The idea of "building up a sense of inevitability", as a strategy for developing and promoting a large-scale technological project with significant national implications, was introduced by Yaakov Garb in his account of the planning and building of the Trans-Israel Highway (Garb 2004). It goes without saying, that the Highway and WEIZAC are two very different kinds of projects in terms of scope, costs, political and social implications, and technological complexity and innovation. Certainly, they cannot be explained or comprehended under identical historiographical categories. Still, we find the term "sense of inevitability" to be very suggestive as a metaphor of the strategy followed by Pekeris in the way to the realization of his project.

[2]SCM held on Jan. 11, 1952 (CPA).

The original version of this chapter was revised: The correction to this chapter is available at https://doi.org/10.1007/978-3-030-25734-7_5

© The Author(s), under exclusive licence to Springer Nature Switzerland AG 2019
L. Corry and R. Leviathan, *WEIZAC: An Israeli Pioneering Adventure in Electronic Computing (1945–1963)*, SpringerBriefs in History of Science and Technology, https://doi.org/10.1007/978-3-030-25734-7_3

Pekeris not only paved the way to a smooth approval of the project and created this sense of inevitability. He actually made the project viable by mobilizing financial resources, transferring knowledge, and recruiting the necessary professional personnel. In order to have his plans materialize he put together a team of talented engineers and scientists, some of whom came to Israel especially for the task. He found ways to acquire the necessary electronic equipment, some of which was hardly available in most countries of the world, and much less so in the entire Middle East. Particular difficulties arose in the attempt to purchase a Magnetic Core Memory to serve as the main storage unit in the machine. None of these tasks could be deemed easy to achieve, but Pekeris's incredible resourcefulness, combined with the complex network of connections that he had developed throughout the years, proved to be crucial and to allow for the successful realization of the project.

In this chapter we explain how Pekeris orchestrated his efforts, how the decision was taken at the Institute to build the machine, how Pekeris led its actual construction, and, finally, how he succeeded in making WEIZAC operational within a relatively short time.

3.1 The Institutional Challenge: Promoting an Unlikely Technological Project

When he joined the planning committee of WIS in 1946, Pekeris was guided by a clear vision of the "dire need of modernization both with regard to equipment as well as methods of manufacture"[3] in Palestine. The Weizmann Institute—and his own project for building an electronic computer within that institution—would play a leading role in the way to make this vision come true. Pekeris argued that building an electronic computer would be an important project for the institute as well as "a typical example in which the commercial importance of skill and ingenuity is conspicuous."[4] For him the electronic computer was an instrument to achieve technological progress and to enhance the engineering skills of the local engineers, as well as a tool to advance mathematics and science. He conveyed his ideas to Weizmann and he seems to have convinced him from quite early on. Thus, on November 1947, Weizmann wrote to one of his friends, Max Brailowsky: "Pekeris is confident that one could eventually build an electronic computer in Rehovot."[5]

Nevertheless, Pekeris's ideas were not unanimously welcomed and the decision to go ahead with the project at WIS met with many hurdles. In a memorandum ordered by Bergmann in 1946 with recommendations, based on national economic priorities, for constructing a Technical Physics group in Rehovot, the construction

---

[3]Pekeris to Bergmann, Mar. 9, 1946 (WIA 3–22–46).

[4]"The Following Examples", in the Pekeris-Bergmann correspondence file (WIA 24–74–6).

[5]Weizmann to Max Brailowsky, Nov. 26, 1947 (LPCW, Series A, Vol. 23, letter 46).

of an electronic computer was not even listed.[6] This group was meant to include "a number of young scientists… who are also capable or devising and building new types of apparatus [sic] in industry which … would be extremely suitable for Palestine."[7] The construction of "electronic computers" (the plural form appearing in the original document) is mentioned only in a second report on priorities prepared at roughly the same time, in relation with the Technical Physics Group and the industrial products that would be derived from these projects. In a meeting of members of the planning committee in 1946, with the WIS architect Pekeris requested more space for the DAM, since:

> [Pekeris] is now definitely planning to undertake experimental work, particularly in the field of the construction of Electronic Computing machines.[8]

Pekeris invested considerable effort among the influential people at WIS aimed at gathering the necessary funds. His arguments often involved more wishful thinking and visionary sights than actual facts. Interesting evidence in this regard appears in a letter of December 1946 to Getzoff, a staff member of the American Committee of WIS, who acted as its "field director."[9] As usual, Pekeris described the design of his intended electronics lab and repeated the arguments about the economic importance of the project. But he also added now additional interesting arguments. The electronic computer—he explained to Getzoff—was the result of the scientific effort that surrounded World War II. Its computing power would enable the actual solution of important scientific problems, which required many months of intense calculations by human teams and which were thus prohibitive. He explained his intention to build a computer based on the latest innovations in the field, originating from leading institutions such as Harvard and Princeton, where prototypes already were in place. Scientists who were expected to join soon the institute's staff at Rehovot, he added, were already on the computer teams of these research centers. The actual situation, however, was more complicated than what Getzoff could gather from Pekeris's description. Electronic computer technology was at the time far from being mature in any sense, and its viability in commercial projects was still far into the horizon.[10] In addition, the candidates with computer

---

[6]"Memorandum of Recommendations for the Technical Physics Laboratory," Oct. 28, 1946 (WIA 23–74–6). The signature in the bottom of this draft document (of which we did not final a final version) is not clear. However, the document opens with the words: "Dear Bergmann, as agreed upon I am submitting recommendation on …".

[7]Bergmann, in the Annual Meeting of the American Committee of Weizmann Institute of Science, Inc., Dec. 10, 1947 (WIA 17–2016–5).

[8]Stern, "Minutes of Part of Planning Committee and Others October 15, 1946, at Polytechnic Institute of Brooklyn" (WIA 6–74–23).

[9]Weisgal, "Report to the Board of Directors of the American Committee and to Committees in Other Countries for the Weizmann Institute for the Period of 1944–1949," 37 (WIA).

[10]The first electronic computer used in a commercial context was the LEO I (Lyons electronic office I), which was modelled after the EDSAC and became operational in 1951 in the UK. See Land (2014).

experience that Pekeris had in mind were, as we will see below, potential candidates at best, and as a matter of fact, none of them finally came to WIS.

And yet, given Pekeris's adherence to his vision and the strong will to bring this project into fruition, the idea of the electronic computer gradually established itself among the decision makers, albeit always with hesitations. In a letter to David Ginsburg (1913–2010), an American political activist in the Jewish Agency and a personal advisor to Weizmann, Bergmann described the kind of research that could to be carried out in the five departments[11] of the planned institute. For the DAM he mentioned research in electronics aimed at "contributing to the efforts now made to construct an 'electronic brain' which will solve complicated problems mechanically."[12] In the 1946 memo sent to Getzoff, in which the laboratories of the new institute are detailed, Bergmann wrote about the laboratory for electronics:

> In this laboratory some work will be done on the theory and the fundamentals of the application of electronic devices required in many of the most modern apparatus [sic], including the electronic computer, which has also been called the mathematical brain.[13]

Bergmann was slow to commit himself to the project, but he promised that the decision on the research plans was the exclusive authority of each department head:

> It was understood that the laboratories at the disposal of the department of applied mathematics will be devoted to electronic research and to research in geophysics. It is not the intention of the Institute to carry out production or manufacturing of any apparatus, for which the Technical Physics Group is intended. Here again, the decision rests with the Head of the Departments, who will take into account the general facilities available to all in the new institute.[14]

The correspondence between Pekeris and Bergmann on this matter continued until July 1947, when the Advisory Committee of the DAM was established. At this point, Pekeris asked for the support of Albert Einstein. Einstein agreed to be a member alongside other truly prominent figures such as von Neumann, Robert Oppenheimer (1904–1967), Abraham Pais (1918–2000) and Hanes H. Kramers (1894–1952), but he refused to be the committee chairman. As he wrote in a letter to Pekeris:

> I consented to be a member of both the sponsoring Committee for the Weizmann-Dinner in November and for the Advisory Committee for the Department of Applied Mathematics of the Weizmann Institute in Rehovot. But in no case have I accepted to be chairman. May I ask you to correct this error immediately.[15]

---

[11]The five departments listed in this letter are: Department of Applied Mathematics, Department of Optics, Department of Biophysics, Department of Isotope Research, Department of High-Polymer Research.

[12]Bergmann to David Ginsberg, Nov. 21, 1946 (WIA 5–74–42).

[13]Bergmann, "A memorandum to Getzoff," Dec. 2, 1946 (WIA 6–74–24).

[14]Bergmann to Pekeris, Dec. 11, 1946 (WIA 3–96–25).

[15]Einstein to Pekeris 1947. The letter is not dated. A copy of the original letter is in (WIA 3–96–45).

Obviously, Pekeris was trying to gather support by drawing into the project the most scintillating names available, but at the same time he did not refrain from attracting some less-known rising stars, or at least to make believe that he could be able to draft them. This was the case with Zvi Lipkin (1921–2015) and John Blatt (1921–1990), two younger, successful scientists that Pekeris was sure in 1946 to be able to recruit for the project, but who never actually joined. Lipkin[16] was then at the Radiation Lab at MIT, whereas Blatt was a mathematical physicist at Cornell, with working experience at RCA (at the time one of largest manufacturers of consumer electronics) (Franklin 2001). Whether or not the two ever intended to join, Pekeris made sure to boast such names as potential meaningful recruits. He also stressed their connections in the industry as valuable assets that would putatively contribute to the success of the project. In a letter to Bergmann, Pekeris described the two of them as "*Anshei Shlomeinu*" (Hebrew expression for "people of our own camp") that he would be meeting soon at Princeton. The letter vividly reflects the feeling of how Pekeris conveyed his enthusiasm to those he intended to persuade. Thus he wrote:

> I plan to get in touch with them and suggest that they take an active interest in the construction of the "super" electrical computing machine now being carried on by Zworykin.[17] My aim is to have these fellows design and construct such a machine for our department of Applied Mathematics. It is my hope that after the development of the first model at R.C.A. the construction of additional units will be relatively inexpensive (i.e. in terms of the present estimated cost of 1/4 to 1/2 million$).[18]

Still, Bergmann proved to be a hard cookie. He did not think that an advisory committee was a very useful idea, as he believed that "it would look much like window-dressing."[19] Nevertheless, Einstein did make an effort to meet the Swiss chemist Camille Dreyfus (1878–1956), who established a foundation as a memorial to his brother Henry, in order to raise money for the DAM. He wrote to Dreyfus:

> I would be very glad of an opportunity to discuss with you the newly founded Department of Applied Mathematics at the Weizmann Institute at Rehovot, and its research program. It you could come to Princeton for a visit, it would be welcomed not only by me but also by Dr. Oppenheimer and others here who are also deeply interested in the project.[20]

These efforts did not bear fruit, as Dreyfus canceled the meeting due to bad weather conditions.[21] Pekeris recalled several years later that the committee did recommend building a computer in Rehovot:

---

[16]Harry (Zvi) Lipkin (Weizmann Compass). Feb. 4, 2016. https://www.weizmann.ac.il/WeizmannCompass/sections/people-behind-the-science/harry-zvi-lipkin (accessed Jan. 14, 2018).

[17]Vladimir Zworykin (1888–982) led in the 1930s the development of the electronic television system at RCA. In the 1940s the company was also involved in the development of automatic computers.

[18]Pekeris to Bergmann, Mar. 9, 1946 (WIA 5–74–26).

[19]Bergmann to Pekeris, Jul. 30, 1947 (CPA Fondiller file).

[20]Einstein to Camille Dreyfus, 1946–1947. The letter is not dated, (WIA 45–3–96).

[21]Camille Dreyfus to Einstein, Dec. 30 1947 (WIA 45–3–96).

Einstein was hesitant. This was the very beginning of the era of electronic computers. They were in a highly experimental stage, and very expensive. But von Neumann's persuasive powers won Einstein over and the advisory committee unanimously approved the plan (Pekeris 1964).

In a letter to Eugene Black, President of the International Bank for Reconstruction and Development, Pekeris recounted that von Neumann had persuaded Einstein by claiming that he, Pekeris, needed the best computer available: "Oppenheimer last Fall said to me: 'wherever you are there will have to be the best of computers.' These, I remember, were the very words by which von Neumann won over Einstein and Oppenheimer for the original WEIZAC project."[22] And thus, in 1947, the advisory committee reached the decision to support the project (Estrin 1991; Pekeris 1964).

As the new Institute continued to be built in Rehovot, Bergmann allocated in the general budget one fifth of the $250,000 to each of the five departments for purchasing equipment.[23] Pekeris insisted on securing this budget for funding the computer project, which he planned anyway to delay for a while. During the years that elapsed until the project materialized, the money was diverted for other purposes. It is interesting to notice that we found no evidence in the records indicating that such an amount was ever explicitly dedicated to the project. And yet, Pekeris continued to refer to this amount, claiming that Weizmann had allocated $50,000 to build a computer even though he did not use it immediately.[24] This became for him a main motto in the campaign for promoting the computer in years to come.

All the while, Pekeris made sure to stay updated with the state of the art technology in computers. Thus, for instance, he participated in one of the first conferences about computers ever—"A Symposium on Large-Scale Digital Calculating Machinery"—that took place in Harvard on January 1947.[25] He sent the Symposium program to Bergmann, for showing it to Weizmann, and wrote:

I am enclosing herewith the program of a scientific symposium to be held in Harvard, on the occasion of the opening of the Computation Laboratory. I thought you might want to show this to the Chief. I could get additional copies of the program if they should prove useful to Mr. Getzoff.[26]

By the time he arrived in Rehovot, Pekeris referred to the decision to build the computer as a done deal, and indeed as a project that was inevitable. He was successful in consistently developing a discourse that was gradually adopted at WIS at large, in which the electronic computer appeared as the only preferred solution to many of the main scientific problems that were on the agenda of the various

---

[22]Pekeris to Eugene Black, May 11, 1960 (CPA).

[23]Bergmann to Pekeris, Dec. 11, 1946 (WIA 3–96–25); Bergmann to Pekeris, Jul. 30, 1947 (CPA —Fondiller file).

[24]Pekeris to Frei, Feb. 7, 1952 (CPA); (Pekeris 1964).

[25]Pekeris' name appears in the "Members of the Symposium" list in (*Proceedings of the Symposium on Large-Scale Digital Calculating Machinery* 1948. Reprinted 1985, xxv).

[26]Pekeris to Bergmann, Dec. 11, 1946 (WIA 3–96–25).

departments. He continually repeated the story of the $50,000 budget that had putatively been allocated by Weizmann himself for the construction of a computer and made sure to present the project efforts as if they were already under way. When the time came for the scientific committee to decide on the project, the discussion was—as we will now see in the remaining pages of this section—not about the necessity of a computer but rather on how to finance it.

In its first years, after the arrival of Pekeris, the DAM provided services to other departments of WIS as well as initiating its own research projects. Pekeris made sure to use this situation as a stepping-stone on the way to solidifying the sense of inevitability for the computer project. For example, in the area of "chemistry of food," the help of the DAM staff was needed in order to statistically compute "the effects produced by an enzyme on glutamic acid," which takes part in the biosynthesis of proteins. Joseph Gillis, from the DAM, cooperated in this project with the physical chemist, Israel Dostrovsky (1918–2010), and the biologist, Aharon Katchalsky (1913–1972).[27] Gillis worked out the mathematical theory based on a model proposed by the other scientists, who also experimentally verified the theory.[28]

The DAM was also engaged in establishing a laboratory devoted to the geophysical survey of the country and to support ongoing efforts to search for water and oil. Likewise, the department team was also occupied with the analytical work needed to prepare problems for the electronic computer. Pekeris used the scientific report to bring up again the electronic computer subject. He reported on the completion of the construction of electronic computers at some leading research laboratories in the USA and the UK. He also stressed that this was an opportunity to use such machines to solve problems in physics and chemistry, since an automated machine would decrease the computation time by a factor of 10^5.[29]

In a meeting of the Scientific Committee in May 1950, Bergmann asked whether the DAM could help Gerhard Schmidt (1919–1971), the founder of X-ray crystallography research at WIS, to speed up the heavy calculations involved in his crystallography project. Schmidt, an Oxford graduate, was a prominent figure in the field, which was at the time a rather innovative one that the Institute leadership was keen on encouraging. This was the kind of situation that Pekeris never missed in his campaign on behalf of the computer project. He thus replied sharply, and presented as definitive facts some ideas that were somewhat speculative and processes that were yet to materialize:

> An electronic computing machine could speed up such calculations, and it would be desirable for the institute to become seriously interested in building such a device. From the work done at Princeton we can conclude that building such a device would cost about $

[27]WIS, "Scientific Activity Report 1949" (WIA).

[28]"Scientific Activity of the Department of Applied Mathematics of the Weizmann Institute of Science during the Period of September 1951 to September 1953" (CPA).

[29]Pekeris, "A Report on Research Work in the Applied Mathematics Department 1950" (CPA) (Hebrew); WIS, "Scientific Activity Report 1949" (WIA).

50,000. Dr. Frei was admitted to Princeton for a year, and there is hope that after his return he will be able to work out a plan for building the above-mentioned machine.[30]

Two months later, Pekeris submitted a work program of the DAM for 1951 to Bergmann. The estimated price for the project was $100,000, whereas in the previous citation the estimation was only at $50,000.

Already in 1949, Pekeris started to design solutions for problems in quantum mechanics and other related subjects, and prepared them for being solved with an advanced electronic computer.[31] One of these concerned the properties of Helium II. In a letter to von Neumann he listed a system of equations that represent the problem and wrote:

> I venture to say that most of the outstanding problems in quantum mechanics which have been given up as "too difficult" will on close inspection turn out to have complexities of the above nature which though more involved than the ordinary run of equations, are still readily amenable to the electronic computer.[32]

Von Neumann answered:

> The differential equations which you give as determining the problem of Helium II do not look vicious. How bad are they from the numerical point of view? They should certainly not be bad for high-speed computer. Quite apart from more ambitious devices, the ENIAC or the SSEC,[33] or the Harvard Mark II or III should be sufficient here ... It would certainly be good to discuss these and other similar problems personally when you are here.[34]

The work plan for the years 1951–1955, that Pekeris submitted to Bergmann in 1950, comprised requirements for manpower qualified for running an electronic computer and for developing new calculation methods suitable for use with it.[35] At this point—we should stress once again—no decision had yet been taken to go on with the project. In a meeting of the Scientific Committee it was noted that the electronic computer "that will be built by Schmidt and Dr. Frei," will greatly increase the speed and efficiency of the work in the X-ray laboratory.[36] Pekeris rejected any alternative plan that was suggested to solve computational problems. For example, Fred Hirshfeld, a junior researcher in Schmidt's group, enquired Frei about materials needed to build an analog network for solving simultaneous linear equations (30 × 30). Schmidt had intended to build such a network at the estimated cost of 3000 $. In his response to Hirshfeld, Frei relied on the expert opinion of

---

[30]SCM help on May 9, 1950 (CPA). (Hebrew). For details on the role of Dr. Efraim Frei in the project, see below.

[31]Pekeris, A Report on Research Work in the Applied Mathematics Department 1950 (CPA). (Hebrew).

[32]Pekeris to von Neumann, Feb. 15, 1950 (WIA 3–96–72).

[33]IBM's Selective Sequence Electronic Calculator.

[34]Von Neumann to Pekeris, Feb. 23, 1950 (WIA 3–96–72).

[35]Pekeris, "A Work Plan of the Applied Mathematics Department for the Years 1951–1954," 1950 (CPA).

[36]SCM held on Nov. 22, 1950 (WIA 1–6–76).

Estrin as well as that of Herman Goldstine (1913–2004)—one of the leading figures in electronic computing at the time and von Neumann's second in command in the IAS machine project—who, according to Frei, "considers it a very bad idea". Frei concluded that there is no "valid reason" for building the analog computer.[37] The leasing of a calculating machine from IBM, which was suggested as a possible way to address Schmidt computation problems, was also rejected off-hand.[38] At this time, Pekeris already got a "go ahead signal" from Frei concerning the current status of the IAS computer. Frei anticipated that the IAS computer would be completed in about three months, even though the completion date is "officially one week."[39] Later that year, Pekeris rejected another alternative solution to WIS computational concerns: being part of a computer center planned to be built in Rome by UNESCO.[40]

Pekeris also made sure in various opportunities that the necessary budget be allocated for the project. Thus, for instance, in a meeting of the Scientific Committee in 1951, in whose report we read:

> Dr. Pekeris also raised the question of the electronic computer. Originally promised to the Department of Applied Mathematics and now also needed for Dr. Schmidt in connection with the crystallographic work. Dr. Schmidt had come to the conclusion that the I.B.M. machine, originally planned to assist in the computing of the crystallographic work, was no longer considered a desirable acquisition, and he would be prepared to forgo the proposed expenditure of $7000 a year, and allocate this sum for the next two years ($14,000) towards building of a computer...[41]

And yet, the most influential person at the Institute, when it came to matters of raising and allocating funds for an expensive project of the scale of an electronic computer, was Weisgal. Pekeris provided him, the "layman",[42] as Weisgal called himself, with an appealing argument:

> Your story that it would take nine years to make the computations of the existing problems, as against six months with the use of the computer, is a very telling argument, and I think I might be able to sell it on these grounds.[43]

Weisgal, together with Pekeris and Frei, had to decide about the specifications of the computer. When considering the possibility on settling on a machine modeled after the IAS computer, they estimated that "even if the money and equipment is available for the new machine ... it could not be completed in less than five years'

---

[37]Frei to Pekeris, Sep. 26, 1951 (CPA).

[38]SCM held on Nov. 8, 1951 (CPA). See quotation bellow.

[39]Frei to Pekeris, Sep. 26, 1951 (CPA). For more on Frei and the IAS machine, see below.

[40]Pekeris to Frei, Jan. 8, 1952 (CPA).

[41]SCM held on Nov. 8, 1951 (CPA).

[42]Weisgal, "Report by the Chairman of the Executive Council for the Period from November 2, 1949 to Jun. 30, 1952" (draft) (WIA 12–90–17). In this document, Weisgal distinguished between "we, who labor for the Institute as laymen" and the scientists.

[43]Weisgal to Pekeris, Nov. 12, 1951 (WIA 3–91–102).

time."[44] Pekeris proposed alternative solutions. One was buying the computer of the English company, Ferranti, a Manchester-based company that was the first to produce commercially available, general-purpose computers; another was building a cheap computer. Frei thought that all existing machines, other than the IAS machine, were already obsolete by then, and that a cheap computer would be to the IAS machine "like the Rehovot Railway station to the Pennsylvania Station in New York."[45] Eventually, it was Pekeris how decided on the IAS model, following Weisgal policy: "Nothing but the best for the Weizmann Institute."[46]

On January 11, 1952, the Scientific Committee of WIS discussed the budget of the DAM and, as part of this discussion, the chairman suggested to decide unequivocally whether the Institute should commit itself to the computer construction project. In that meeting, Pekeris pointed out that the electronic computer was an old project of WIS for which $50,000 had been definitely allocated with the consent of Weizmann. He repeated his claim that the money was deposited in the bank. As we mentioned earlier, however, no such allocation of money was indeed effected. One way or another, the committee did approve unanimously, and Weisgal was entrusted with ensuring the initial budget of $13,000, a sum which was about 10% of the estimated cost.[47] A formal decision and a clearly authorized budget were of course necessary conditions for making Pekeris's dream come true. But more than that was needed and Pekeris was clearly aware of that. His efforts were directed now towards putting together a team of skilled and enthusiastic engineers and scientists, to ensuring the location and infrastructure for the machine within WIS, and to making sure that the necessary components for the machine could be purchased and shipped to Rehovot in a timely fashion. These aspects of the project are described in what follows.

3.2   The Human Challenge: Putting Together a Team

Pekeris understood that his project for an electronic calculator in Rehovot depended crucially on putting together a qualified team of enthusiastic partners. As already mentioned, some of the staff would be recruited, according to his initial vision, from the teams currently working in similar projects at the leading institutions in the USA and the UK. Some of his choices materialized but, to be sure, not all of them.

Morris Rubinoff (1917–2004) is a remarkable example of a candidate that Pekeris worked hard to gain over for the project, but was unsuccessful at that. At the beginning of 1946, when Rubinoff completed his doctorate in Canada, he offered

---

[44]Weisgal to Pekeris, Nov. 12, 1951 (WIA 3–96–102).
[45]Frei to Pekeris, Jan. 18, 1951 (CPA).
[46]Pekeris to Weisgal, Nov. 20, 1951 (WIA 3–96–102), Frei to Pekeris, Dec. 5, 1951 (CPA).
[47]SCM held on Jan. 11, 1952 (CPA).

Pekeris his candidacy for the DAM[48] and planned to stay one year in the United States to expand his horizons before coming to WIS. His plans matched well the timetable for setting up the DAM in Rehovot, which was about to be inaugurated a year later. Rubinoff came to Harvard University to work in the physics department and, encouraged by Pekeris, he divided his time between the physics department and the computer laboratory under the direction of Howard Aiken (1900–1973), the leading figure in the Harvard computer project (Aspray 2000a, 51). In Aiken's laboratory, Rubinoff participated in the development of Mark III, the first electronic computer in the laboratory to operate with vacuum tubes instead of relays.[49]

Pekeris invited Rubinoff, as a computer expert, to attend the planning committee meeting of WIS at the Polytechnic Institute of Brooklyn in October 1946, despite the fact that Rubinoff had joined Harvard only a few months earlier. At the meeting, Pekeris requested space at WIS for the computer, based on data gathered from Harvard and MIT computers. Pekeris introduced Rubinoff as a future partner and as expert in building electronic computers. Rubinoff informed that the budget for the computer currently built at Harvard was of about $500,000, and confirmed Pekeris's requirement for lab space.[50]

Rubinoff was on the list of new researchers of the DAM and planned to start at WIS on July 1948.[51] Eventually, he did not come to Israel, but he kept in touch with the Institute's staff, came to visit[52] and in 1960 he also advised the computer purchase team of the Israel Defense Forces.[53]

Pekeris also looked for suitable candidates in the UK. On November 1946, he wrote to Joseph Gillis (1911–1993) a mathematician who had worked as cryptographer during the war at Bletchley Park.[54] The two had met for the first time in Cambridge, while Pekeris visited there in the mid-1930s. After the war, Gillis moved to Queen's University at Belfast and made plans to immigrate to Palestine. He met Bergmann while the latter was in London, they talked about the possibility of joining the DAM at Rehovot and Bergmann referred Gillis to Pekeris. Pekeris wrote him specifically about the project:

---

[48]Rubinoff to Pekeris, Mar. 11, 1946 (WIA 5–74–26).

[49]Rubinoff, Interview by Richard R. Mertz, May 17, 1971, (SOVA,1969–1973, 1977).

[50]Stern, "Minutes of Part of Planning Committee and Others October 15, 1946, at Polytechnic Institute of Brooklyn" (WIA 6–74–23).

[51]"Annual Meeting of the American Committee of Weizmann Institute of Science, Inc., Dec. 10 1947" (WIA 17–2016–5), Weisgal to Pekeris, Nov. 13, 1947 (WIA 6–74–24). Also, Weizmann requested immigration certificates for Rubinoff and his wife in a letter to Alan Cunningham, the High Commissioner of Palestine, Jan. 6, 1948 (CWA2806–18A).

[52]Pekeris to Estrin, Apr. 7, 1958 (WIA 12–90–79).

[53]Mordechai Kikion (founder and first commander of IDF Center of Computing and Information Systems), "A Report on the Visit to the US 25.4.1960–30.6.1960" (IDFA 5–327–1965).

[54]It seems unlikely, however, that Pekeris was aware of Gillis' cryptographic work during the war.

It may interest you to know that Dr. Booth[55] and Prof. Newman of Manchester are coming to Princeton this Winter to study with von Neumann various phases of the design and application of the electronic computer. We are actively pushing a project to build an electronic computer for the Dept. of Applied Mathematics in Rehovot.[56]

In 1948 Gillis settled in Rehovot and joined the DAM.[57] One of his assignments was managing the gravity aspect of the geophysical survey of Israel. After Bergman left, in addition to his research in mathematics, Gillis became the academic secretary of the WIS. During the construction of WEIZAC, in 1954–55, Gillis spent a sabbatical year on the IAS computer project. Upon his return, Pekeris left for sabbatical and Gillis replaced him as head of the department. As part of this role, he was responsible for allocating the WEIZAC computing resources to external users, in the first year of its operation. Gillis used WEIZAC in various projects, but was concerned about the enthusiasm of the department's young people with the electronic computer and worked to "… keep in front of the younger people the fact that computing is neither the whole of mathematics or [sic] a substitute for it."[58]

Ephraim Frei (1912–2006) was another important figure at the WIS computing community, as we have already seen above. A physicist who had arrived in Palestine from Vienna in 1940, Frei worked at the Hebrew University until 1945. During the 1948 war, Frei was deeply involved in HEMED, the Science Corps of Israel, and later was a main figure in the electronics department of the ministry of defense.[59] Frei was considered by many to be the "number one" Israeli expert in radio technologies and to have "created almost all of the electronics in Israel."[60] Furthermore, he "personally, knew almost everyone working in the field in Israel" (Estrin 1991). As already mentioned, Frei spent some time in Princeton around 1950, and since his arrival there he was involved in the IAS computer project. He also gradually became involved in the task of acquiring equipment and components for the future project, and gained much experience in the field,[61] thus became the right person to lead this side of the project.

In early 1948, von Neumann suggested to Pekeris that one of the members of WIS should apply for a scholarship at the IAS at Princeton.[62] Pekeris was unhesitant in choosing Frei for applying, and he was indeed chosen for the

---

[55]Andrew Donald Booth (1918–2009) was an English computing pioneer that worked on the IAS computer project (Computer Pioneers—Andrew Donald Booth, http://history.computer.org/pioneers/booth-ad.html, accessed Aug. 7, 2018.

[56]Pekeris to Gillis, Nov. 17, 1946 (WIA 6–74–24).

[57]Before joining the DAM, during the 1948 war, Gillis served in the IDF Intelligence unit. See Nahem Ilan in Pe'amim 122–123, Yad-Ben_Zvi. https://www.ybz.org.il/_Uploads/dbsAttachedFiles/122–123_Nahem_Dmuyut_193–212_13.5.10.pdf.

[58]Gillis to Pekeris, Oct. 21, 1956 (CPA).

[59]"Weizmann Award Laureate 1957," DAVAR, Jun. 12, 1957, 4. (Hebrew). See also Bacharach (2009, 64–66).

[60](SCM held on Feb. 14, 1951.) (Hebrew).

[61]Pekeris and Frei, Pekeris–Frei correspondence, Dec. 22, 1950–Sep. 25 1951 (CPA).

[62]von Neumann to Pekeris, Jul. 28, 1949 (CPA).

scholarship. Frei spent the years 1950–52 in the computer research team of the IAS, just on time to join the last stages of computer project there that was still in progress. Pekeris reported to the Executive Committee about this invitation and insisted that as soon as the computer would be completed in Princeton, a similar computer would start to be built at WIS.[63] He added: "That may cost you people some money, but anyhow they are going to get it free at the expense of the United States Government through a fellowship."

While Frei was still at Princeton, Pekeris tried to secure a permanent appointment for him at WIS, so he would work on the DAM computer project. However, HEMED electronics department had to release him from duty. Bergmann, then still the scientific director of WIS, also served as scientific adviser to David Ben-Gurion (1886–1973), who asked him to undertake the mission of defining the role of science for the defense and economic development of the newly established state (Jensen et al. 2011). When the Scientific Committee discussed Frei's appointment, Bergmann represented HEMED, and claimed that "the electronic group of HEMED cannot give up on Frei's work."[64] Bergmann proposed a replacement for Frei, which Pekeris objected, because he did not believe that the proposed candidate "could do the work for which Frei is required, such as an electronic computer [sic]."[65] Thus, Frei, not receiving an official appointment from WIS, decided, in full agreement with Pekeris, to accept the proposal to extend his stay at Princeton as a member of the computer group, under Julian Bigelow's leadership.[66]

On May 1951, as already indicated, Bergmann left his position as WIS scientific director, following differences of opinion with Weizmann concerning the participation of WIS in military activity. At the end of that month, the Ministry of Defense's scientific department agreed to release Frei so he could join WIS in 1952.[67]

While Pekeris was fighting in Israel for Frei's appointment, Frei expanded his knowledge on computers. He worked with Bigelow on electronic computing research topics such as the theory of adders. His status as a researcher gave him access to the computer that was being built, but spared him the pressure of the project. In addition, he studied the reports of the development of the Whirlwind computer at MIT (Redmond and Smith 1980), which at that time was declassified. He also participated in a conference on vacuum tubes for computers, in Atlantic City.[68]

As a member of the computer team at the IAS, Frei was able to update Pekeris on the status of the project, whose schedule continuously slipped.[69] Frei knew the team members well and therefore was able to identify the right person to run the

---

[63]"Annual Meeting of the American Committee of Weizmann Institute of Science, Inc. Dec. 8, 1948" (WIA 6–74–24).

[64]SCM held on Mar. 27, 1951 (CPA). (Hebrew).

[65]SCM held on Mar. 16, 1951 (CPA). (Hebrew).

[66]Frei to Pekeris, May 17, 1951 (CPA).

[67]SCM held on May 28, 1951 (WIA 6–76–2).

[68]Frei to Pekeris, Jan. 18, 1951 (CPA).

[69]Frei to Pekeris, Jan. 18, 1951 (CPA).

Fig. 3.1 Engineering group and leaders of the computer project at the Institute for Advanced Study, Princeton, including two WEIZAC team membres, Frei and Estrin. From left to right, Gordon Kent, Ephraim Frei, Gerald Estrin, Lewis Strauss, Robert Oppenheimer, Richard Melville, Julian Bigelow, Norman Emslie, James Pomerene, Hewitt Crane, John von Neumann. From the Shelby White and Leon Levy Archives Center, Institute for Advanced Study, Princeton, NJ, USA. *Photo Credit* Alan Richards

project in Israel. To do so, he had to prevent Pekeris from giving this role to Bigelow. Bigelow's work on the IAS project had made him the most prominent figure in the field, but according to Frei, while Bigelow was an excellent researcher and engineer, he was a bad manager. He wrote to Pekeris that members of the team accused Bigelow for the recurring delays in the project's completion.[70]

Frei's choice fell upon Gerald Estrin who would become the chief engineer of WEIZAC. Estrin was born in New York on September 9, 1921 to a Jewish family. In 1951 he won his Ph.D. in electrical engineering at the University of Wisconsin. A short time before the completion of his doctorate he and his wife, Thelma Estrin (1924–2014), herself also with a Ph.D. in electrical engineering from Wisconsin, joined the IAS computer project as research engineers. At Princeton they met Pekeris and his wife, Leah, as well as Frei and his wife Yael (Fig. 3.1). These acquaintances led to the offer made to Estrin to join the team in Rehovot and become its

---

[70]Frei to Pekeris, Dec. 5, 1951 (CPA). According to Aspray, "Bigelow's perfectionist tendencies slowed progress so much that von Neumann had to replace him with James Pomerene in the midst of construction" (Aspray 2000a, 187).

technological leader. At the end of 1953, the Estrin's left for Israel. While Gerald ran the WEIZAC project, Thelma participated in the planning until the birth of their child. In April 1955, they returned to the United States, but maintained personal and professional contact with the WIS people. Later on, Estrin was appointed to the Institute's Board of Directors. Upon his return to the United States, Estrin joined UCLA as a faculty member and was one of the founders of the Department of Computer Science there. He was the head of this department during 1979–1982 and 1985–1988. One of his main contributions to the field of computer architecture is the development of reconfigurable computing. This concept led to the development of programmable computer chips (Lee 2012; Estrin 1991).

During 1954–55, before WEIZAC became operational, Pekeris started to look for mathematicians with programming experience. Obviously, they were very rare at the time. The first candidate was Philip (Pinchas) Rabinowitz (1926–2006), who got his Ph.D. degree in mathematics from University of Pennsylvania in 1951 and started to work at the Computation Laboratory of the National Bureau of Standards, Washington, DC. There he acquired his programming experience on the SEAC computer.[71] On October 1954, Pekeris expressed his will to engage Rabinowitz as a mathematician at WIS.[72] Rabinowitz "made aliya" and soon became the WEIZAC chief programmer. Rabinowitz's contribution to the development of the software for WEIZAC cannot be overestimated. He prepared the software infrastructure of WEIZAC, which includes a mathematics library and many other utilities. In the summer of 1955, Rabinowitz gave the first programming course in Israel.[73]

The Estrins arrived in Rehovot by the end of 1953, with slight delay due to the birth of their first daughter in February that year. Their arrival marked the beginning of the practical, engineering stage of the WEIZAC project. Alongside Gerald, Thelma Estrin was a leading moving force, participating in the design of the machine all along her stay in Israel, except for one month, when she gave birth to her second child (Nebeker 1993). With the help of Frei, the first staff members were recruited. Additional team members that joined in 1954 included Micha Kedem (1931–2001), Aviezri Fraenkel (b. 1929) and Zvi Riesel (1922–2002). Fraenkel had just graduated from the Technion as an electrical engineer and then completed his army service before joining the team.[74] Skilled manpower in electronics was hardly available in Israel at the time, and much less so the academically trained. Zvi Riesel was among the few such. He was born in 1922 in the city of Halle, in Germany. At the age of 15, he immigrated to Israel

---

[71]SEAC (Standards Eastern Automatic Computer) was an early computer developed at the National Bureau of Standards. It became operational on May 1950. See Kirsch (1998). At roughly the same time, also the SWAC (Standards Western Automatic Computer) became operational in the West Coast (Corry 2008, 40–48). They were the fastest operational computing machines before the IAS computer became operational one year later. Both machines played an important role in the widespread adoption of electronic computing in the USA, both in science and in administration.

[72]SCM held on Oct. 21, 1954 (CPA).

[73]"Contribution of the Weizmann Institute to the Advanced of the Science of Electronic Computing Jun. 5,1961" (CPA). See also Davis and Fraenkel (2007).

[74]For more information on Fraenkel see Scheinerman and Simpson (2001).

with his family. Due to financial difficulties, he dropped out from high school and went in the direction of vocational training in electronics, which he pursued at a local lab of Philips, the Dutch radio company.[75] During the Second World War, he served for three years in the British Army. He thereafter served in the Israeli Defence Forces (IDF), a time during which he also completed his academic studies (1947–1950). Upon being released from duty, in 1954, he joined the WEIZAC team.

The WEIZAC people compensated for lack of proper training and experience, by strenuous work and quick acquisition of the necessary knowledge, as evident from the letter from a Telemeter engineer who came to the streets to install the magnetic memory who wrote:

> May I state that I was greatly impressed with the caliber of people associated with the computer. Zwi Reisel and Co. are first rate computer people. Their spirit and know how greatly shortened the time required to get the computer and memory on the air.[76]

In 1955, following Estrin's return to the USA, Riesel became the team's manager for the next nine years. He also participated in the design of the first and second generations of the GOLEM computers, which were later built at WIS (see below). In 1966, he was appointed associate professor and over the following years, he advised doctoral students in the field of computer design at both the Weizmann Institute and the Technion. It should be pointed out that in this early period of WIS, special appointments of this kind existed whereby distinguished engineers became faculty members in parallel to the standard academic procedures of the institution. While Fraenkel went on to pursue a distinguished career as a mathematician at WIS, Kedem and Riesel, together with other younger members that joined the team, eventually became prominent forerunners of the computers establishment in Israel, both in academy and in the industry.

All in all, the team involved in the initial development of WEIZAC was relatively small, and it was only later, in the framework of the GOLEM and GOLEM B projects, that broader cadres of talented young personnel were incorporated as budding members of the ever-growing circle of pioneers in the computer community of the young State of Israel.

3.3   The Material Challenge: The Lab, the Components, the Budget

Reaching a formal, positive decision to go ahead with the project at WIS and putting together the right team were indeed two important, necessary conditions for WEIZAC to materialize. The account presented in the foregoing passages

---

[75]The Hebrew daily press provides plenty of evidence about the active presence of this company in the main cities of the Yishuv, at least since the 1930s. See, e.g., *Haboker* Feb. 24, 1938; *Hayarden*, March 05, 1935; *Al Hamishmar,* Aug. 4, 1944. This information was retrieved from the Historical Jewish Press site, at the National Library, Jerusalem: http://web.nli.org.il/sites/JPress/English/Pages/default.aspx (accessed Aug. 17, 2018).

[76]Milton Rosenberg to Pekeris, Oct. 4, 1956 (CPA).

underscores the significant difficulties encountered in the way to achieving them. Further substantial challenges, however, still lay ahead, particularly concerning the ability to provide for the material aspects of the project. Among the main challenges, the following three were the most pressing ones: (1) setting up an electronics laboratory, (2) obtaining all the necessary electronic components, and, most crucially, (3) purchasing the fast memory that would allow for the speedy functioning of the computer.

The material challenge was truly daunting. In the early 1950s, WIS was far from having the required financial reserves. According to Bloch, the administrative director of WIS, "the institute in fact had no regular income and lived from hand to mouth."[77] The committee's decision of January 1952 contemplated a budget of $10,000 in the first year for the project, and probably, at best, the same amount over the next two years. Most of this budget was intended for importing components, and Getzoff emphasized that there would be no additional funding for the project from the Institute's sources. It was thus clear to Pekeris and to his team that the lion's share of the standard components should be obtained, in good Zionist spirit, as contributions from the Institute's friends who manufactured or marketed electronic equipment.[78]

But additional difficulties arose from the rather backward technological environment and poor overall infrastructure in Israel at the time. Estrin described vividly in his memories "the bare shelves and dusty windows of electronic equipment stores in Tel Aviv" (Estrin 1991):

> Essentially no materials were available in Israel and we needed vacuum tubes, tube sockets, resistors, wire, solder, sheet and bulk aluminum, sheet copper, sheet insulating materials, potentiometers, meters, tube testers, resistance testers, hardware, soldering irons, and miscellaneous tools.

The electric power supply in Israel, in addition, was "very poor" with frequent interruptions, and the Institute had to buy its own stand-by generator.[79] Also the cooling system was not fully reliable still by 1956.[80] But precisely because of these considerable challenges, the existing leadership behind the project proved to have been an ideal choice. On the one hand, Weisgal as fund-raiser, with the assistance of the American Committee of WIS and its director Getzoff. On the other hand, Frei and Estrin on the technical side. Together, they were indeed able to ensure that the material conditions for executing the project would become available, as we describe in detail now.

Let us start with the construction of the lab. In the early days of WIS, a "tinker shop" served the technical needs of the various departments. The construction of a computer involved a great deal of mechanical work and required continued and

---

[77]SCM held on Mar. 25, 1954 (CPA).

[78]Pekeris to Bigelow, Jan. 15, 1952 (CPA); Weisgal to Pekeris Dec. 27, 1951 (CPA).

[79]Pekeris to Frei, Aug 27, 1952 (CPA).

[80]And just to provide the right context for these kinds of difficulties it may be worth mentioning that according to the testimony of R. Narasimhan (1926–2007), head of the TIFRAC project in India, the activation of their computer was delayed for almost a year, due to the lack of air-conditioning facilities (Narasimhan 1960).

exclusive use of the devices, as well as of space. The tinker shop could not provide any of this.[81] In the middle of 1952, however, a solution was found arising from an unexpected direction. In his search for funding for the computer project, Weisgal contacted Benjamin Abrams (1893–1967), a Romanian-born American business-man, who in 1922 had founded the first company that sold radio-phonographs in the USA, the Emerson Radio & Phonograph Corporation. Abrams, as it happened, was not impressed by the proposal to build a computer.[82] Instead, he offered to finance the construction of an electronics department at WIS. Frei prepared a proposal for such a department, which included a team of 15 people and a budget of $350,000. Abrams seems to have been after a much bigger project, however, and he was willing to provide a budget three to four times larger than this amount.[83] In November 1954, the Benjamin Abrams Electronics Lab was inaugurated in Rehovot, and in its first year, the computer project occupied the laboratory almost entirely.[84] Later, the lab was devoted to other projects in electronics under the direction of Frei, and eventually, towards the completion of the project, the com-puter and its team moved to the Applied Mathematics Department.

Parallel with the efforts to have a laboratory in place, a second fundamental task of the project involved the purchasing of the electronic components. The efforts in this direction had begun way before the scientific committee reached a positive decision in 1952 and it continued in various ways until the magnetic core memory was finally purchased in 1956. Initially, Pekeris had hoped to start the project by the end of 1952, after Frei's return to Israel from Princeton, and to complete the project by the end of 1954. The amount and variety of components required to build first-generation elec-tronic computers was significant. The electronic circuits of the IAS computer contained about 3000 vacuum tubes of five different types. The cost of the electronic equipment was of tens of thousands of dollars and most of it was not available in Israel.

The required materials comprised, in the first place, standard components such as copper wires, standard tubes, resistors and the like (Table 3.1).[85]

Some of the early efforts of Weisgal and Frei to gather such components as contributions from Jews in the USA yield modest, yet important success. Thus, for instance, Frei was able to persuade Harry Cohn, the owner of one of the largest stores for electronic equipment surpluses in Detroit, to provide for the Institute some of the electronic equipment needed to build the computer. Cohn took upon himself the task and organized a group of local manufacturers and distributors of electronic com-ponents who helped obtaining, usually free of charge, the needed parts.[86] In April

---

[81]Frei to Pekeris, May 29, 1952 (CPA).

[82]Frei to Pekeris, Apr. 10, 1952 (CPA).

[83]See Footnote 81.

[84]WIS, "Scientific Activity Report 1954" (WIA).

[85]This list is a combined summary of information appearing in two documents: (1) "List of parts required for the construction of electronic computer", May 1st, 1953 (CPA); (2) "Tentative list of materials for electronic computer", 1952 (CPA).

[86]Getzoff to Bloch, Jun. 4, 1952 (CPA).

Table 3.1 Weizac components—partial list (1952/1953)

Component	Amount required for the computer	Amount of spare parts (expected rejects and testing)
Tubes:		
6J6[a]	1500	3000
Various	1000	1000
2C51	300	300
Resistors:		
Wirewound	100	100
Small-carbon	7000	3000
Others	100	100
Condensers:		
Mica paper	2000	2000
Electrolytic	200	200
Sockets:		
Standard tubes	2800	200
Switches	100	50
Indicators:		
Pilot lamps	25	50
Neon	250	250
Copper wire (12 different types):		
Round wires (3 types)	4800 ft	
Stranded insulation	4000 ft	
Tinned copper	4000 ft	
8 colors (3 types)	20,000 ft	
4 colors (2 types)	750 ft	
Formvar copper (2 types)	600 ft	
I/O equipment:		
Page printer	1	
Perforators	2	
Reperforator	1	
Tape reader		
Fax printer		

[a]According to Estrin (1991), the J6J tubes were eventually replaced by tubes of a different type

1952, Cohn promised Frei that a considerable portion of the tubes, capacitors and other components were already available.[87] Pekeris, however, was not satisfied and he asked Weisgal to become more actively involved:

---

[87]See Footnote 81.

The collection of electronic parts, which is being shouldered by a single person, Mr. H. Cohen [sic] of Detroit, is not proceeding as well as we had expected. This is a rather big project. I would like to ask you to take the matter in your own hands.[88]

Still, the process of gathering the standard components continued all the while. In May 1952, Hershel Radio Co., another Detroit-based company, supplied hundreds of condensers and tubes, as well as 3400 tubes sockets.[89] Soon thereafter, in July, Harry Cohn supplied all the necessary resistors. The amount of standard tubes needed for a computer was of around 2000, but reliability levels were a real problem, and hence components needed to be selected from larger stocks so as to meet stringent quality constrains. Accordingly, the actual number of components needed was 2000 tubes and 12,000 resistors.

The exacting task of putting together the more standard components was essentially completed by July of 1952, and it was much more than an empty lip-service when the booklet of the annual conference of the American Committee of WIS, published in 1954, stated:

The American Committee for the Weizmann Institute of Science tenders its grateful acknowledgment to the many manufacturers of electronics equipment who have most generously contributed their products towards the construction of our electronic computer.[90]

Obtaining the non-standard components, however, proved to be a much more difficult challenge. Particularly difficult was the issue of the special memory tubes, which Frei planned to get from RCA as a contribution. More specifically, Weisgal and Frei had pinned hopes on David Sarnoff (1891–1971), an American Jewish pioneer in radio and television who ran RCA for many years, as the right person to help them obtaining them. As early as 1949, Sarnoff had expressed his willingness to build a small electronics factory in Israel.[91] In 1952 Eliezer Kaplan (1891–1952,) Israel's first Minister of Finance and later on Deputy Prime Minister, invited Sarnoff to come to Israel, hoping that he would build a local vacuum tube factory.[92] Sarnoff visited Israel in 1952, met with Ben-Gurion and offered to support Israeli television, but Ben-Gurion doubted Israel's financial ability to establish such a project.[93] During his trip, Sarnoff also visited WIS and met Pekeris. Sarnoff promised to assist both in supplying the rare tubes and in terms of ease of payment.[94] Pekeris cynically wrote to Frei:

If Weisgal succeeds in getting this favor out of Sarnoff, he will be so happy that I think he will be willing to put out the money out of his own pocket in order to exploit [it].[95]

---

[88]Pekeris to Weisgal, May 27, 1952 (CPA).

[89]Hershel Radio Co., invoice, May 5, 1952 (CPA).

[90]"Three Anniversaries," Dec. 2, 1954 (WIA subject file: computers 2).

[91]Ben-Gurion, Diaries—full diary Jul. 1, 1949 (BGA).

[92]Ben-Gurion, Diaries—full diary Feb. 4, 1952 (BGA).

[93]Ben-Gurion, Diaries—full diary Jul. 26, 1952 (BGA).

[94]Pekeris to Frei, Jul. 29, 1952 (CPA).

[95]Pekeris to Frei, Aug 13, 1952 (CPA).

The components that Pekeris and Frei hoped to receive from RCA were forty tubes of 1024 bits for the computer's primary memory. The expected cost of this crucial component, commercially called "Selectron", was $56,000. This type of memory had been initially designed for the IAS computer, but RCA was behind schedule and according to Bigelow, in the spring of 1948 there were doubts about RCA's ability to support a parallel memory of 4096 words with their Selectron (Bigelow 1980, 303). Therefore, von Neumann and Bigelow decided to develop, in-house, a parallel memory of only 1024 words. Eventually, the Selectron was used only for the JOHNNIAC computer, which became operational in 1953 and its capacity was 256 words only.[96] The WEIZAC team also clearly understood that the memory of their machine would need to rely on a different type of technology. We expand on this interesting issue in the next section.

Another important, non-standard component was the I/O unit. In January 1952, just before the Scientific Committee reached the decision to go ahead with the project, Weisgal traveled to London and one of his tasks was to purchase I/O devices at Ferranti.[97] After the committee's approval and the allocation of a preliminary budget for the first year, Pekeris made sure that the order for the devices that Weisgal acquired in London would not be reduced from this budget:

> The above appropriation is exclusive of whatever Mr. Weisgal may make from Ferranti. I would like to keep that angle under the hat, since otherwise Bloch will deduce it from the $10,000.[98]

Another way to get free parts was to collect them from out-of-use stocks. Frei and Getzoff met a senior IBM executive who was involved in the development of IBM first production computer IBM 701.[99] They tried to get through him the power supply of an old computer as a donation. The IBM people, however, eventually dismissed the request and reused the old power supply for experimental purposes.[100]

WIS also tried to obtain financial support from external sources, such as international organizations and the Israeli government. For example, on October 1951, the Scientific Committee requested a grant of $50,000 from UNESCO for the computer project.[101] WIS did not get the grant, because at a conference held by UNESCO in Paris, late 1951, it was decided to establish a calculation center in Rome. Though Israel became a member of this center,[102] Pekeris, with the support

---

[96] As explained in *The RAND* Apr. 1954, *Digital Computer Newsletter*, Vol. 6(2), 3–4. See also Davis and Fraenkel (2007).

[97] Pekeris to Frei, Jan. 8, 1952 (CPA). On the Ferranti computers see Napper (2000), Blachman (1953).

[98] Pekeris to Frei, Jan. 15, 1952 (CPA).

[99] IBM Archives: IBM 701, www-03.ibm.com/ibm/history/exhibits/701/701_intro.html (accessed Jan. 15, 2018).

[100] Frei to Pekeris, Jun. 13, 1952 (CPA).

[101] SCM held on Oct. 30, 1951 (CPA).

[102] UNESCO, "Convention for the Establishment of the Internation Computing Center," Jan. 7, 1952 (CPA).

of Schmidt, believed that an electronic computer in Rome would not fulfill WIS's computing needs.[103] Another attempt was to get funds from the Ford Foundation. In this case, the request was for "Application of Modern Automatic Computing Machinery to the Interpretation of Geophysical Measurements," as part of a Geophysical Survey of Israel.[104] This project was a cooperation of several local research institutions. Pekeris planned to use the funding for purchasing the magnetic core memory, but eventually the Israel Foundations Trustees[105] did not transfer the money to WIS.[106]

Perhaps one of the most remarkable points concerning the financial side of the WEIZAC project was that, contrary to many other contemporary, similar projects, it was based exclusively on private funding. On various occasions, Bergmann and members of the scientific committee suggested that the military be asked to contribute to the funding effort.[107] We have found no evidence, however, that such suggestions ever materialized.[108] It is possible that the relevant technological units at IDF came to appreciate the significance of the project only much later (see below §3.6).

Just before leaving Israel to return to the USA in March of 1955, Estrin presented at the meeting of the Scientific Committee, a rough estimation of the expected costs of the electronic computer project at WIS. His estimations were as follows[109]:

- foreign purchases: $48,000,
- local purchases, wages, workshop spending, outsourcing and administration: $24,000,[110]
- magnetic core memory: between $140,000 and $220,000.[111]

Eventually, however, the total cost of the project ended up being about $350,000,[112] namely more than one third of the total annual budget of WIS for 1954 (i.e., $1,000,000).[113]

---

[103]Pekeris to Frei, Jan. 8, 1952 (CPA).

[104]Israel Trustees Foundation, "First Comprehensive Report." June 1955, 161–162. (ISA-most-ScienceResearchboard-00070xh).

[105]The Israel Foundations Trustees was a committee established in the USA in order to recruit foreign funds for scientific research in Israel.

[106]SCM held on Feb. 3, 1957 (WIA 1976–6–8).

[107]Pekeris to Frei, Sep. 25, 1951 (CPA); SCM held on Jan. 11, 1952 (CPA).

[108]A similar conclusion about the lack of involvement of IDF in the financial side of the WEIZAC project is found in (Krauss 1997). Krauss based his research on interviews conducted with the leading physicists of WIS during the 1950s and 1960s.

[109]Estrin, "Report on Electronic Computer Project," Mar. 21, 1955 (CPA).

[110]The report states 36,000 Israeli Pounds. Here we indicate a rough estimation, due to the difference between official and formal exchange rate.

[111]We found contradictory estimations for this expected cost.

[112]SCM held on Feb. 3, 1957 (CPA).

[113]"Scientific Activity Report 1954" (WIA); Rate conversion based on electronic resources of the Bank of Israel: http://www.boi.org.il/he/Markets/ForeignCurrencyMarket/Pages/shearim48–77. aspx (accessed Jun. 4, 2018).

3.4   The Magnetic Core Memory Challenge

One of the central technological challenges that arose in the construction of all early computers centered on was the issue of the primary fast memory. As the use of electronic computers continued to grow, the need for increasing amounts of fast memories became increasingly pressing. Existing memories did not have sufficient capacity, they were slow, and did not allow random access or were very expensive and had reliability issues (Ceruzzi 2003, 44–45, 49). This was a critical problem, which threatened to stop the advance of the technology, by causing what the historian of technology Tom Hughes described as a "reverse salient." In his influential book on the development of electricity grids at the turn of the 20th century (1983, 14–15), Hughes adopted the military term—"reverse salient"—as a metaphor to describe the situation in which a specific instance of technological advance can create local withdrawals due to critical problems. Hughes claimed that in such situations, researchers, inventors and engineers become very actively involved in trying to find solutions, and indeed optimal solutions tend to appear in several research centers simultaneously. Surely, this explanation applies very fittingly to the case of computing technologies as well. Concerning the issue of the fast memories, such optimal solutions were embodied in the magnetic core memory technology, developed in several research centers simultaneously, at the beginning of the 1950s. The list of inventors involved includes the likes of Jay W. Forrester at MIT, Jan A. Rajchman at RCA, An Wang at Harvard, and others (Pugh 1984; Eckert 1953).

As already mentioned, the question of the fast memory was also a crucial, initially unresolved one in the case of WEIZAC. Indeed, the issue arose as early as 1952, right after the Scientific Committee had decided to approve the project. Frei was rather discouraged at the time because of the difficulties encountered in achieving the components, particularly from IBM and RCA, as explained above. In August 1952 he even suggested that it would be better to build a smaller computer.[114] But in conversations with Bigelow new ideas arose, as the latter suggested avoiding the use of the problematic RCA-type of memory initially intended for use with the IAS machine. Instead, he proposed to build a magnetic drum, a technology originally developed back in the 1930s, that had been broadly adopted since the mid-1940s (Rubens 1999). Frei objected, arguing that a parallel machine with a magnetic drum would be inefficient and that even if multi-heads drum were used, the machine would be too slow and the coding very complicated.[115] Instead, Frei proposed that, in addition to the drum, they would build a 256-digit memory, based on Williams-Kilburn Tube cathode-ray technology, the first known type of random access memory. Williams tubes, first patented in the United Kingdom in 1946, afforded storage capacities

---

[114]Frei to Pekeris, Sep. 02,1952 (CPA).

[115]A parallel machine, like the IAS computer, performs an arithmetic operation on all bits simultaneously, while a serial machine performs the operation bit after bit, like the EDVAC, and the operand can be fetched from the memory bit after bit. Therefore, a parallel machine performs arithmetic operations much faster, as long as the memory access time is fast enough.

ranging from 1024 to 2048 bits (Pomerene 1972, 978). These tubes were very difficult to obtain at the time, and thus only a smaller-size memory could be currently guaranteed. Frei believed, however, that in no more than two years, RCA or other manufacturers would overcome the technological problem involved in the production of memory tubes.[116]

Bigelow wrote to Pekeris expressing his objections to the use of an electrostatic memory (Williams' tubes). At IAS, a memory of this type was functioning well at that time, but Bigelow believed that in Israel, where the available electronic equipment was rather scarce and primitive, it would be difficult to build, maintain and operate such a memory. Instead of the electrostatic memory, Bigelow suggested three levels of memory: (1) a number of registers from conventional vacuum tubes; (2) a small, fast magnetic drum with an average access time of one millisecond; (3) a large magnetic drum to serve as secondary memory. He estimated that the speed of this set of memories would be "something like 1/2 or 1/4 but more than 1/10 that of our present machine."[117]

As already said, one explicit goal of Pekeris's project for an electronic computer in Rehovot was to develop practical competence in engineering in Israel. The project was not so much one of striving for innovation, but rather one of being as close as possible to the IAS computer, which was the design model of WEIZAC:

> I had vowed to hold design changes to an absolute minimum in order to have a reasonable chance to finish a working machine in Rehovot by 1955. During our travels I had been intrigued by Wilkes' microprogramming concepts but withstood the temptation and allowed only those changes that were driven by reliability considerations... (Estrin 1991).[118]

Still, the issue of supporting access to a larger memory required a major change in the computer architecture. The size of the IAS computer memory was 1024 words, despite of the original intention of the designers to support 4096. The reason to the reduced memory was the technical inability, at the time, to come up with a fast memory of the desired size. The WEIZAC designers decided to support a memory of 4096 words. This decision implied the need to redesign the main control unit in order to provide memory locations with lengthier access addresses. As it happened, Pekeris was able to take full benefit of this size when he decided to tackle an important open physical problem, namely that of finding a solution for the wave equations for the two-electron atom. He addressed this problem in terms of a numerical solution reached by a memory- and computation-intensive approach (Pekeris 1958). This work became Pekeris' most cited one, and it required the full exhaustion of the WEIZAC memory.

During the two years that elapsed between the decision to build the computer and its actual construction, important developments took place in memory technology. This was also the time when Estrin became aware of the possibility of

---

[116]Frei to Pekeris, Mar. 24, 1952 (CPA).

[117]Bigelow to Pekeris, Apr. 22, 1952 (CPA).

[118]Maurice Wilkes (1913–2010) was a highly distinguished British computer pioneer. He designed the EDSAC, which was model after the EDVAC, but was operational already in 1949. In 1951 he published a seminal paper about microprogramming (Wilkes 1951).

Fig. 3.2 MNEMOTRON—core memory of WEIZAC. Courtesy of Ruth Reisel & Neomi Yaron

purchasing a commercial magnetic-core memory. The prospective manufacturer was the International Telemeter Corporation, a California-based company intent on introducing at the time an innovative kind of PayTV service. This company was established by Paramount Pictures Corporation in 1949, under the leadership of Adolph Zukor and Barney Balaban and was commercially motivated by the fear of losing movie audiences to the new TV technologies. They hired computer engineers and television experts as part of their efforts to adapt their movie distribution systems to the new technological reality (Eisner 1981). As part of their R&D efforts, Telemeter began to produce 4096-word magnetic-core memories for commercial purposes. Weisgal spoke to Balaban to ensure the earliest possible supply of the memory and of favorable terms of payment: a significant discount in the price and distribution of the payment over a number of years, by selling long-term notes of the American Committee of WIS.[119] Estrin worried about the high price of the device, and he confidentially asked Telemeter's VP for Engineering, Louis N. Ridenour, to sell to WIS the components and the technical documentation, so as to allow the WEIZAC team to build the memory in-house.[120] Ridenour refused the request.[121]

---

[119]Weisgal to Hammer (telegram), Sep. 12, 1954 (WIA 12–90–63).

[120]Estrin to Ridenour, Nov. 12, 1954 (CPA).

[121]Ridenour to Estrin, Nov. 26, 1954 (CPA).

On April 1954, the first memory that Telemeter produced, the Mnemotron, was shipped to the JOHNNIAC project. They supplied two additional memories to Argonne National Laboratory and to the Ballistic Research Laboratory in Aberdeen. WEIZAC got the fourth unit[122] and on September 14, 1956 (Fig. 3.2), the memory was put into operation.[123] On September 1956, Gillis wrote to Pekeris "The new memory is now connected up … The speed of the machine is quite frightening."[124] At this point, all the necessary pieces for completing the WEIZAC project were already in place.

3.5   The Implementation Challenge: Building and Operating WEIZAC

The first months after the arrival of Thelma and Gerald Estrin in Rehovot were dedicated to set the project ready for work. A team was recruited, the electronic parts were ordered, and the construction equipment was put in place. Estrin defined the technical specification and requirements, and trained the new staff in the theory of building computers. He delivered lectures in English and Riesel—who had served in the British army for several years, and was fluent in English—translated simultaneously into Hebrew.[125]

The WEIZAC team had the full set of drawings of the IAS computer, but like all the first-generation computers, the WIS computer had to be custom-tailored: while the general logic was identical to that of the original, some unique features had to supplement it. The modifications of the technical specifications were intended for increasing the reliability of the computer and were derived from lessons learned with the IAS computer. They also had to satisfy Pekeris's demand for as much memory as possible. In the Scientific Activity Report of 1954 the first activities were summarized as follows:

> The first six months were spent gathering materials, staff and space; producing the stamped parts required for the basic chassis assemblies; redesigning where necessary; making the first high speed memory investigations; conducting a weekly seminar and beginning the selection of the 2000 vacuum tubes and 12,000 resistors.[126]

Fraenkel re-designed the "main control" (Fig. 3.3) unit in order to enable an increase in the size of the address space by a factor of four (Estrin 1991), as mentioned above. During the day Fraenkel was at work with the design and in the

---

[122]Raymond Stuart-Williams, International Telemeter Corporation, to Pekeris, Nov. 29, 1955 (CPA).

[123]"Scientific Activity Report, 1956–1957" (WIA). We have no detailed information about how Weisgal finally stroke the deal.

[124]Gillis to Pekeris, Sep. 13, 1956 (CPA).

[125]"Richard Solomon, Estrin, Weizac and Golem Pioneers". Video interview (partially supported by the American Committee for the Weizmann Institute of Science 1983 —private copy).

[126]"Scientific Activity Report 1954" (WIA).

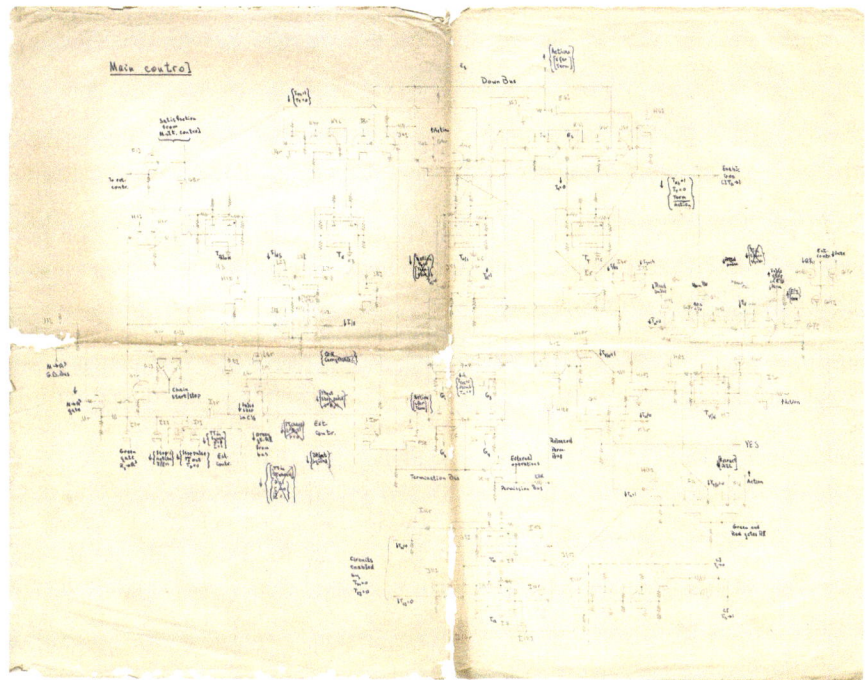

Fig. 3.3 WEIZAC main control. Courtesy of Ruth Reisel & Neomi Yaron

evenings, Estrin and Riesel examined the design and discussed it with him.[127] Thelma Estrin helped design one of the crucial components of the machine, the so-called "logical adder." Her innovative design replaced the one used in the IAS machine, known as the "Kirchoff adder."[128]

Towards the end of 1954, Estrin and his team worked around the clock to advance the project as much as possible before Estrin's expected leave by early April 1955. In March 1955, after nine months of intense work, the main units were almost complete. The power supplies, the air-conditioning system and most parts of the computer itself were also installed. The I/O devices had not yet been completed. On March 20, 1955, the computer was connected to the power supply in the presence of guests invited by Pekeris. According to Fraenkel,[129] they forgot to make sure that there were no short-circuits in the system, and when the computer was turned on, a capacitor exploded loudly, and sparks were scattered in the air.

Estrin summarized the event in these words:

---

[127]Aviezri Fraenkel, personal correspondence, Jul. 2, 2013; Segel Lee, "Conversation with Riesel Zvi, Feb. 18, 1987" (HMF).

[128]"Scientific Activity Report 1954" (WIA). See also Estrin (1991).

[129]Fraenkel, private interviews with Leviathan, 2009–2013.

> At 10 p.m. yesterday, March 20th, the central computer was shocked into a new phase of its existence when the power supplies were connected to the machine and turned on. It wasn't a completely uneventful ceremony—there were no speeches—only one loud burst of a condenser blowing up and showering sparks all over.[130]

The next four months were devoted to testing, while the memory of the magnetic drum was completed and, together with the Teletype equipment, connected to the computer. After Estrin departure, Riesel became the chief engineer of the group. Estrin mentioned in his report that "he is fully capable of carrying on from here" and added, "You are fortunate in no longer being dependent on the importation of foreign know-how."[131]

In June 1955, all the arithmetical operations worked properly, and the development of the control unit of the Ferranti paper tape reader was in full swing. The output equipment, Flexowriter, had arrived and the planning of its control unit was in progress. Estrin, who had returned to the US, investigated the market of input-output equipment looking for faster units.[132]

In October 1955, the machine ran its first program, with the magnetic drum serving as the primary memory. According to the "Report on Computer Users," of November 1955, the first programs computed square roots and differential equations. During this month, the total run-time was of 61:4 h, of which 47:35 h were devoted to code checking and 14:35 h to production.[133] Interestingly, as time went by, program check-time decreased significantly, relative to production time.[134] Obviously, the programs were written in machine code. As described in the "Notes on Coding the WEIZAC":

> When all instructions are arranged in the right sequences, they are copied to a coding sheet and explicit address filled in... The last step is to translate the symbolic code into machine code reading for making up a tape.[135]

When the machine became operational, there was a need to use a formal name in scientific articles. In July 1956, in a letter to Estrin,[136] Pekeris suggested to officially choose the name that was already accepted and currently in use at the department:

WEIZAC—Weizmann Automatic Computer

In April 1955, the following column appeared in the *Digital Computer Newsletter* of the Office of Naval Research:

---

[130]Estrin, "Report on Electronic Computer Project," Mar. 21, 1955 (CPA).

[131]Estrin, "Report on Electronic Computer Project," Mar. 21, 1955 (CPA).

[132]"State of Computer Jun. 5, 1955" (CPA).

[133]The term "production" was commonly used to describe the run time of the correct program.

[134]In Pekeris' file, "Computer Reports, December 1955" (WIA 3–96–38).

[135]"Notes on Coding the WEIZAC," Sep.1960 (https://archive.org/details/notes_on_coding_for_weizac_sep60).

[136]Pekeris to Estrin, Jul. 5, 1956 (CPA).

The geographical frontier of electronic computing will be extended with the completion of a modern high-speed electronic computer at the Weizmann Institute of Science, in Rehovot, Israel.

Construction of the newest member of the family of IAS machines was begun in June 1954. The Central computer is scheduled to begin tests during March 1955, using a drum memory constructed at Rehovoth. A core memory system of 4096 words produced by International Telemeter Corporation will be installed later in the year.

With the exception of the high-speed memory, corresponding changes in the control and the replacement of Princeton's Kirchoff adder by a logical adder, the Israeli machine follows the design of the computer at Princeton.[137]

Public news about the machine appeared for the first time in an Israeli newspaper at the end of 1955 under the title: "Electronic Brain Activated at the Weizmann Institute."[138] Since the beginning of 1958, the name WEIZAC appeared in various scientific publications, as well as in the Digital Computer Newsletter.

During the operational period of WEIZAC, the market of peripheral equipment for computers developed significantly. Consequently, hardware adjustments were made to allow the connection of new equipment to the computer. The three most important such improvements were the following:

- **A magnetic tape storage system** with high capacity and medium access time was built in 1957. Special machine instructions controlled this system. It became apparent that the use of magnetic tapes significantly increased the range of problems that can be solved with the help of WEIZAC. Hence, it was decided to optimize the system by compressing the stored data using the quick shift operation of WEIZAC.[139]
- **A system to enable offline printing directly from the magnetic tape** was designed and built in 1957–1958. The output was written by the computer to the magnetic tape while the program was running. With the new system, data was printed directly from the magnetic tape without tying down the computer. The device was built entirely with transistors and was completed in the spring of 1959.[140]
- **Two new magnetic-core memories** were received as a gift from Telemeter Magnetic in 1960.[141] They were originally meant for the ERMA, which was a pioneering machine used for data-processing in the banking industry.[142] The memory word-size of the ERMA was smaller than that of WEIZAC and hence it

[137]*Digital Computer Newsletter*, Office of Naval Research—Mathematical Sciences Division, Vol. 7 (2).

[138]"'Electronic Brain' activated at the Weizmann Institute," *Al Hamishmar,*Oct. 24, 1955. (Hebrew).

[139]Scientific activity report 1958–1959 (WIA).

[140]Scientific Activity Report 1958–1959 (WIA).

[141]In 1956, the magnetic core memory business of International Telemeter was moved to its new subsidiary Telemeter Magnetics.

[142]Scientific Activity Report 1960–1961 (WIA).

Fig. 3.4 Dr. Gerald Estrin at work on WEIZAC (1954). Courtesy of Werner and Anat Braun. *Photo Credit* Werner Braun

was decided to use these memories as buffers to optimize the access to the four existing magnetic tape systems. It was a large project, second only to the WEIZAC project itself, and was led by Irving Weiselman of Telemeter Magnetics, who came to WIS for a lengthy period of time.[143] The project included the design of printed circuit boards and the use of transistors and diodes and brought the engineering team of the department into a new era in electronics (Weiselman and Tomash 1991) (Figs. 3.4, 3.5, 3.6 and 3.7).

WEIZAC was the first in a series of computers built at the DAM in WIS. In its most productive periods, it provided at least 300 and over 600 h of computation a month.[144] It was turned off towards the end of 1963, when a CDC–1604 computer was donated to the Weizmann Institute, as an intermediate solution until the construction of the new computer was completed. Indeed, two years after WEIZAC became operational, Riesel and Pekeris started considering the next computer, the GOLEM (Fig. 3.8). As soon as GOLEM was activated in 1965, an identical machine was built at a much lower cost, due to dramatic decreases in the prices of

---

[143]Report on Activities During the Period of January–August, 1958 (CPA).
[144]Scientific Activity Report 1962–1963 (WIA).

Fig. 3.5   Three technicians working on the WEIZAC. "Each of these Israeli workers in electronics came from a different part of the world. From the left: Shoshana Rosenberg immigrated from Hungary eight years. ago, Ruth Péeri is a third generation, native-born Israeli, who learned electronics during her military service; Yáacov Haassan, from Tunisia, took his degree at the Haifa Institute of Technology." (Quoted from: "Three Anniversaries" (WIA subject file: computers 2), Dec. 2nd,1954. Courtesy of Werner and Anat Braun. *Photo Credit* Werner Braun

the components.[145] At the same time, the design of another computer, GOLEM B, was initiated. This new project, however, encountered unanticipated difficulties, and thus the design of GOLEM B was completed by 1974 and it became fully operational only in 1976.[146] By that time, an IBM-computer based computation center was already in use at the Weizmann Institute and it was thus decided not to build any more computers there.

---

[145]Scientific Activity Report 1965 (WIA).

[146]WIS press release Dec. 1983 (WIA). (Hebrew).

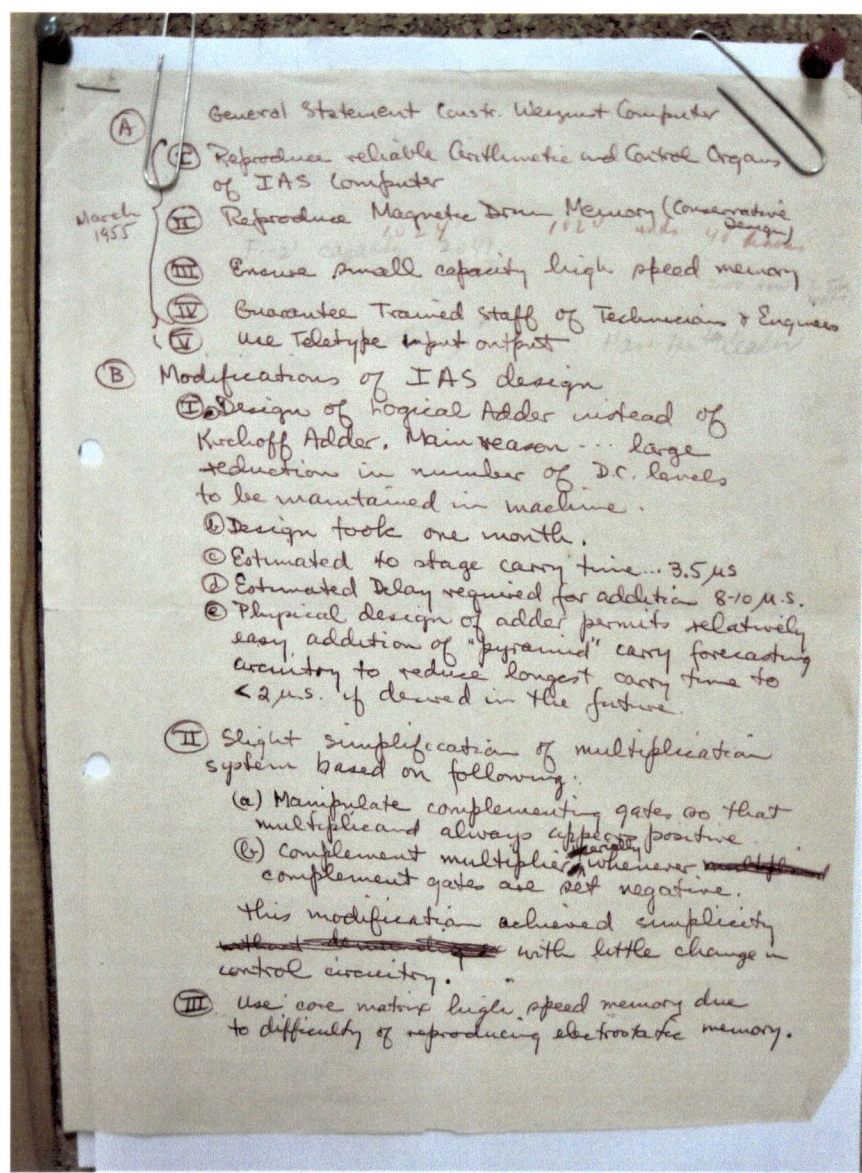

Fig. 3.6 Estrin handwritten "General statement const. Weizmann computer" (1954). Raya Leviathan personal copy

Fig. 3.7 Constructing and checking WEIZAC. Courtesy of Weizmann Institute of Science

3.6   The Scientific Challenge: The Early Impact of WEIZAC

The importance of WEIZAC as a landmark in the early history of computing in Israel and its overall, mid- and long-term influence on the creation of a professional community of programmers and computer scientists in the country have often been pointed out in different contexts (e.g. Ariav and Goodman 1994; Baal-Schem 2007; Breznitz 2002). Remarkably, however, this important historical topic has never been systematically investigated and much less properly documented. It would be beyond the scope of the present study to undertake such a task in detail, and hence we will leave it for a future opportunity. Still, it seems appropriate, by way of concluding our account, to give a brief overview and to indicate some of the main points that need to be considered in doing so.

The first perspective from which to analyze early impact of WEIZAC concerns its actual contribution to the development of science in Israel. For starters, the number of articles whose results were obtained with the support of its

Fig. 3.8 Chaim Pekeris, Pinchas Rabinowitz, Zvi Riesel, and Gerald Estrin with the GOLEM computer (1964). Courtesy of WIS archives. *Photo Credit* Shlomo Ben Zvi. All rights reserved to WIS

computational power is simply startling. Appendices A and B (below) present a summary of relevant information about such publications, which comprise no less than sixty five. Many of these had a significant, direct impact in their respective fields, and they were repeatedly cited throughout the years. For reasons of space, we mention here only some interesting highlights of all of this, and it is interesting to start with some statistics:

- The item with the highest number of citations in this list is an article of 1958 published in the leading journal *Physical Review* by Pekeris himself under the title "Ground State of Two-Electron Atoms" (Pekeris 1958). Over the operational years of WEIZAC, namely between 1956 and 1963, a total of 14,000 articles were published in this journal, and Pekeris's article is ranked in the 80th place, with 800 citations.[147]
- Out of 180 articles published over those years in the same journal by WIS researchers, the two publications with the highest number of citations are by Pekeris, and they are based on calculations performed with WEIZAC.
- Fourteen articles were published over that period of time in the same journal, with results based on computations performed with WEIZAC. The overall

---

[147]This figure was obtained from the journal's website on March 2017. According to Google Scholar, as consulted on Jun. 25, 2018, the actual number was over 1100.

number of citations for these publications is 1885 (i.e., an average of 134 per article).

- The total amount of citations appearing in scientific journals of articles whose calculations were performed with WEIZAC is about 3500.[148]

The importance of Pekeris's 1958 article, and through it, of the early impact of WEIZAC, can be measured not just in terms of its citation ranking, but also in terms of its intrinsic quality as explained at the occasion of the centenary of the journal in 1995. The American Institute of Physics (AIP) published a huge volume entitled *The Physical Review: The First Hundred Years*, comprising about one thousand articles out of about a quarter of a million that were published in the journal since its inception. Pekeris's article was prominently featured in the section devoted to atomic physics, in the following terms:

> Also, in 1958 and 1959 Pekeris—using the Rayleigh-Ritz variational minimum principle and taking advantage of the vast improvements in computing power since the 1930s (though his computers were pitiful compared to today's, of course) —computed the energies of the ground and first excited states of atomic He[149] to a heretofore unheard-of accuracy of about ten significant figures, including relativistic and other corrections; his calculations, wherein he diagonalized 1000 x 1000 matrices, illustrate the growing practice of expanding the sought-for solutions of the Schrödinger equation in function bases chosen less for their expected resemblance to the exact solutions than for their computational convenience (Stroke 1995).

Another Israeli scientist mentioned in the AIP review volume is Giulio Racah (1909–1965), one of the pioneers of physical research in Israel, who arrived from Rome in 1939 after having studied with Enrico Fermi and was appointed in 1949 to the chair of theoretical physics at the Hebrew University. In the 1940s, Racah published a series of papers under the title "Theory of complex spectra," which laid the ground for what is known today as "Racah Algebras" (Stroke 1995; Unna 2000). Racah was one of the most ingenious WEIZAC users. The experience he gained by using WEIZAC during two years of regular and systematic work was summarized in a paper entitled "Use of the WEIZAC in theoretical spectroscopy" (Racah 1958). The paper describes in detail the main strengths of WEIZAC as a tool for scientific research and the basic routines that Racah used in his work. In particular Racah stresses that WEIZAC's memory was "fairly big."

But the impact of WEIZAC was not limited to these important articles or even to these specific fields of research. There can be no doubt that WEIZAC was instrumental in contributing to enhance the international reputation of the Weizmann Institute as a top-notch research center at large. For instance, research conducted by Pekeris together with his research associates Zipora Alterman (1925–1974) and

---

[148]This figure was calculated in 2014 by looking at all articles in Google Scholar in which WEIZAC is mentioned' and by summing up all the amounts indicated in the search results.

[149]"He" stands here for the symbol of Helium.

Hans Jarosch,[150] based on calculations with WEIZAC, received international acknowledgement (e.g. Alterman et al. 1959). The *New York Times* featured an article on this work in August 6, 1960, in the following terms:

> SCIENTISTS DETECT EARTH VIBRATIONS; First Such Observation Is Noted in 4 Laboratories After Chilean Quakes Resonance Is Forecast—Israeli Professor Furnished Calculations …The model of the earth was constructed by Dr. Keith E. Bullen of the University of Sydney Australia, one of the world's foremost authorities on the earth interior, who is in attendance here. The calculation of its resonant vibrations was made by Dr. Pekeris of the Weizmann Institute of Science in Israel.[151]

Many additional important studies, in other research areas such as quantum theory, magnetic materials, X-ray spectroscopy and nuclear magnetic resonance, were supported by calculations performed with WEIZAC. These are summarized in detail in Appendix B below. Over the years, the DAM at WIS was widely acclaimed for its achievements and specifically for the brilliant manner in which it learnt to harness the computational power of WEIZAC to achieve important results in so many disciplines. Interesting evidence of this is found in a statement by Nobel Prize winner, Subrahmanyan Chandrasekhar (1910–1995), who in 1962 wrote:

> I think it can be fairly said that the record of what has been accomplished at the Weizmann Institute with the WEIZAC under the leadership of Professor C. L. Pekeris is unequalled in the world. The uniqueness of this accomplishment derives not so much from the high quality or the large quantity of work that has been done as from the fact that Pekeris and his associates have for the first time used electronic computer for the solution of problems which one could not literally have dreamt of solving before.[152]

A second perspective from which to analyze the actual contribution of WEIZAC to the development of science and technology in Israel concerns the ways in which it influenced the creation of human and physical infrastructures of computer-oriented activities in the country for generations to come. One aspect of this pertains to the establishment of world-class computer science departments in various academic institutions in the country in the decades to come, as well as the early, massive adoption of computer-based techniques in many other disciplines in those same institutions. This aspect requires further investigation focusing on both the development of computer science as a scientific discipline in Israel and elsewhere, and the moot question of whether "early entry" was necessarily "a competitive advantage" in the history of computing (Aspray 2000b). A second aspect pertains to the laying-down of the necessary human and physical infrastructure that would enable the soon-to-come rise of the Israeli high-tech industry and the

---

[150]Alterman started her career at the DAM in WIS, and later established herself at the Department of Environmental Sciences of Tel Aviv University, of which he also became head.

[151]Sullivan, Scientists Detect Earth Vibration. *New York Times* Aug. 6, 1960.

[152]Chandrasekhar, a memorandum on the Application from the Weizmann Institute of Science for a High-Speed Electronic Computer 1962 (CPA).

relatively early, widespread adoption of computers in governmental as well private institutions in the country. These are rather complex topics that we intend to explore in further detail in our forthcoming work, but a hint of it can be grasped by considering the examples mentioned right below.

Throughout its operational years, alongside the extensive, ongoing use of WEIZAC in the research that Pekeris and the staff of the DAM conducted, considerable computer time was also made available to scientists associated with other departments of the Institute as well as with those of other research institutions in the country (such as the Technion and the Hebrew University). Many young students received at the time their basic disciplinary training and at the same time started their professional lives at WIS while relying on WEIZAC (and later on the two GOLEM machines) as a matter of course for the intensive computations required for their research. There is no doubt that the successful computational approach that some of them developed in their work was directly influenced by this part of their training.

Prominent in this regard is the case of X-ray crystallography, which became the fundamental research tool for shaping the seminal ideas concerning biological structure and function at the molecular level (Yonath 2011). It is clear that the availability of powerful electronic computers allowed implementing mathematical methods that played an essential role in solving problems in this area (Hauptman 1990). And as it happened, three Nobel Prize winners whose research focused on this field made their first steps in science at WIS, and they did so while working in projects where the crucial computations were performed with the GOLEM and within the scientific tradition that had arisen, among other things, around WEIZAC. The 2009 Nobel awardee, Ada Yonath, started to work on her Ph.D. thesis in 1965 at WIS. It took one year before she started "to compute."[153] She published in 1969, together with her Ph.D. advisor at WIS, Wolfie Traub, a report on their project to develop a computer-assisted procedure for describing the structure of molecules of collagen (Traub and Yonath 1969).

Two computers that were then operational at WIS, GOLEM and CDC 1604, were of central importance for carrying out this project.[154] They allowed systematizing the methods used for generating structures with helical parameters that were consistent with those observed in the X-ray crystallography experiments (Traub and Yonath 1969, 464). Yonath continued her research towards cracking the ribosome structure, until other, innovative technological advances—including better computational methods and improved computer hardware—allowed the full achievement of the complex crystallographic calculations necessary to do so (Puglisi 2009).

---

[153]Yonath, private email communication to Leviathan, Dec. 21, 2011.

[154]In spring 1960, the Swedish entrepreneur, Axel Lennart Wenner-Gren, supplied WIS with a WEGEMATIC computer, which, however, was significantly inferior to WEIZAC and was hardly ever used.

Also the 2014 Nobel awardees in chemistry, Arieh Warshel and Michael Levitt, started their career at WIS at the end of the 1960s working on questions related to molecular structures, and they did so under the same atmosphere of computation-intensive research that the existence of the machines at Rehovot allowed and, indeed, encouraged. Under the direction of the leading Israeli chemical physicist Shneior Lifson (1914–2001), Warshel and Levitt, were involved in calculations intended to predict the energy of molecular structures and the forces that act upon them as they assume three-dimensional states. Few other places in the world could afford at the time the computing capabilities provided by the GOLEM. They developed a program called CFF, that allowed to fully perform the crucial calculations necessary for understanding the structures of many types of molecules and that became seminal for the entire discipline for years to come.[155]

The interesting issue of the influence of WEIZAC on the eventual development of computer science as a leading scientific discipline at WIS (and in Israel in general) is, as already stressed, far from being self-evident. The first years of "computing" at the DAM were mainly devoted to scientific computation and computer engineering, rather than to theoretical enquiries about computation as such. More generally speaking, over the early 1960s the idea of "computer science" as an autonomous academic discipline was yet to be consolidated. The first "computing" societies were established already in the late 1940's but they mostly dealt with computing machinery: American Institute for Electrical Engineers (AIEE) in 1946 and Eastern Association for Computing Machinery (EACM, later become ACM) in 1947 (Finerman 1968, p. 231; Gupta 2007; Tedre 2014, 33–57). It took ACM almost two decades to come up in 1965 with the "Report of the ACM Curriculum on Computer Science—Preliminary Recommendations." The first department in the USA, named department of "Computer Science" was established only in 1962 in Purdue University.

At WIS, a separate department of computer science was established in 1969 with Shimon Even (1935–2004) as its head. One of the leading first-generation Israeli computer scientists was the 1996 Turing Award winner, Amir Pnueli (1941–2009), and his career is indicative of the impact of WEIZAC on the development of computing, both as an academic discipline and in industry in the country. As a young student at WIS, Pnueli was among the early programmers of WEIZAC. In 1967, he completed his Ph.D. in applied mathematics with Pekeris as advisor. He worked on problems related with oceanographic tides, which required intensive computations (though it is not clear on what machine this was performed). He also participated in preparing the software infrastructure for the GOLEM. In the early 1970s, together with the brothers Ido and Hagi Lachover (both of whom were also involved with the computers of the DAM at WIS); Pnueli founded one of the first

---

[155]https://web.stanford.edu/class/sbio228/public/readings/Introduction_Lecture1/Levitt_NSB_01_History_0501_392.pdf.

software commercial initiatives in Israel, Mini-Systems Ltd.[156] This company developed the basic software for Scitex, one of the first, highly successful local high-tech companies. At the same time, Pnueli started to advise Ph.D. students writing dissertations that fell squarely within the new field of "computer science." Pnueli's first graduate Ph.D., Nissim Francez, completed his dissertation in 1976,[157] and he went on to be among the founders of the CS department at the Technion.

Another important and highly influential initiative that developed around WEIZAC was the Responsa Project. This first-of-its-kind project launched in 1963, which became a landmark in computerized information-retrieval technology, was motivated by a rather unlikely source, namely Fraenkel's long-standing interest in rabbinic literature. As already mentioned, Fraenkel was one of the first WEIZAC designers and later earned a Ph.D. in mathematics in UCLA and developed a distinguished international career at WIS. He was also an observant Jew with vast knowledge of the Jewish sources. The Responsa Literature (שאלות ותשובות שו"ת) was a corpus of Jewish case law that played an important role in the life of communities throughout the centuries, particularly in Europe. It comprised the accumulated collection of answers and rulings given by prominent rabbis in reply to written questions on halakhic issues sent to them by individuals in need for counsel.

Fraenkel came up with the idea that the best way to make these rulings readily available to Jews all-over the world looking for answers to their queries was with the help of a computer-assisted system. He was, no doubt, the ideal person to undertake the development of such a system. Computerized IR (Information Retrieval) systems were first discussed in 1948. Towards the end of the 1950s and during the 1960s the new IR research discipline concentrated on methods to solve two basic problems in the field: indexing and retrieving (Sanderson and Croft 2012, p. 1446). Fraenkel followed the so-called "full-text information retrieval" approach, which was rarely used at the time, and which became popular only in the 1990s, for Web search. Other innovations of the system included the use of morphological analysis[158] and local feedback (Minker 1977; Sanderson and Croft 2012; Xu and Croft 1996). The system, first introduced in 1968, underwent various upgrading stages over the following decades. In 2007, the prestigious Israel Prize was awarded, in the field of Jewish Scholarship, to the Responsa Project. The Prize committee described the project as being the most successful modern tool for leveraging and expanding academic Torah Studies, and for dramatically enhancing the possibilities of continued research in the field.[159]

---

[156]"Amir Pnueli-A.M. Turing Award Laureate"; https://amturing.acm.org/award_winners/pnueli_4725172.cfm, (accessed Jul. 30, 2018).

[157]https://www.genealogy.math.ndsu.nodak.edu/id.php?id=18997.

[158]Morphology deals with word formation, and morphological formation deals with grammatical analysis of a word.

[159]Israel Prize, cms.education.gov.il/EducationCMS/Units/PrasIsrael/Tashsaz/HashutProject/nmk.htm (accessed Jul. 31, 2018).

The impact of WEIZAC and the GOLEM was also felt outside the academic world in Israel. It started by way of courses on the principles of computer architecture and of programming. Such courses were intended for users outside the Weizmann Institute, notably the Israeli Central Bureau of Statistics (ICBS) and the Israel Defense Forces. The ICBS, for example, conducted an industry survey as a case study through which its personnel not only became familiar with the benefits of the electronic computer but also learned how to use it.[160] The IDF, in its turn, used WEIZAC for intelligence tasks and for various weaponry studies. Already in 1955, Bergmann asked Pekeris to allow a representative of the Ministry of Defense Research Division, J. Meiron, to use WEIZAC. Meiron conducted research in optics with the help of the machine in order to solve problems in ray-tracing.[161] Later on, in the early 1960s, the WEIZAC team assisted and consulted the IDF during the procurement process for its first computer, as well as during the subsequent assimilation process.[162] To be sure, by the early 1950s the IDF had become aware of the importance of electronic computers, but to the extent that they were involved in such projects at the time, this was limited to the design and construction of two small analog computers. Munya Mardor (1913–1985), who became head of the Research and Planning Division of the Ministry of Defense in 1952, commented on this matter in relation with the first analog IDF computer, "Mahmad," which became operational in 1956. According to his testimony:

> This computer was still primitive, with "childhood illnesses" similar to those of our first wireless system—great instability, drift, and the need for frequent adjustment and calibration. Nevertheless, even then, it was possible to perform simple simulations … The "Mahmad" was used for R&D until 1958 (Mardor 1981, 184).

A second IDF analog computer, "ITZIG", was developed in the years 1955–59.[163] The real involvement of the IDF with digital computing technology began in early 1957, as Meir (Manny) Lehman (1925–2010) arrived in Israel and joined the Israel Ministry of Defense.[164] Lehman had obtained a degree in mathematics from Imperial College where he also worked on some of the earliest computer projects. He went to do computer research at Ferranti's London laboratory and to complete a

---

[160]"Using the WEIZAC computer for data processing by the Central Bureau of Statistics". Lecture before the Department of Statistics, Jerusalem, Mar. 26, 1958 (CPA).

[161]"Report on computer users", Feb. 1956 (WIA).

[162]Lubin, Ministry of Defence, Ordnance Branch, May 19, 1958 (CPA); Pekeris to Remez, Aug. 30, 1959 (CPA); (Shahar 2002).

[163]E. Erez, "The analog simulator of Machon 3", 1971 (RADC).

[164]Meir Lehman, interview by W. Aspray, Sep. 23, 1993. https://ethw.org/Oral-History:Meir_Lehman (accessed Mar. 16, 2019).

Ph.D. Lehman was the leading force behind the SABRAC computer,[165] which, due to budget shortages, was completed only in 1964.[166]

Beyond the ICBS and the IDF, WEIZAC provided computing services to additional institutions in the country, such as the meteorological service, the Dead Sea Works and the Israel Electric Corporation.[167] Undoubtedly, WEIZAC deserves full credit for the initial stages of the dissemination of computing technology in Israel. The project had both immediate results (such as training, education and other activities) and important long-term effects. It was instrumental in importing computing knowledge to academic and other institutions in Israel, as scientists and engineers arrived in Rehovot from overseas in order to participate in the local activities around WEIZAC and thus they transferred their knowledge to the DAM team members. The presence of a computer at the forefront of the global technology was a source of national pride, which was expressed, for example, in the daily press.[168] During Israel's first decade of independence, a period of immigration absorption and struggling with hostile environment, high-technology projects were exceptional. The contribution of WEIZAC to spreading the idea of innovation in general and to fostering the adoption of computer technology in Israel was also significant. As Yigal Accad, one of WEIZAC programmers, put it: "the fact that after 2000 years, the State of Israel had its homemade electronic computer was for sure an inspiration for the spirit of YES, WE CAN."[169]

The WEIZAC project brought about the recognition of the computer power, the familiarization with programming and the accumulation of experience. Israel's ability to maintain sustained, effective activity around an electronic computer justified a large financial investment in the new technology throughout the early 1960s. WEIZAC was without any doubt the main trigger for the process of adopting the new computer technology in Israel. Pekeris was, of course, greatly satisfied with the achievement, as well as highly optimistic about the future. In 1964, at the occasion of the first conference of Israeli Information Processing Association held at WIS, he expressed himself in prophetic terms, that would soon prove to be realistic, when he declared: "May we merit that every man will be under his vine and under his fig tree and enjoy the brightness of his *private* electronic computer" (Pekeris 1964, emphasis added) (Figs. 3.9, 3.10, 3.11 and 3.12).

---

[165]T. Lamdan, "Developing digital computer in Machon 3", Dec. 1967 (RADC).

[166]Incidentally, WEIZAC was used to simulate the logical design of SABRAC. See Lehman et al. (1963).

[167]"Report on computer users" (1959, 1960) (WIA).

[168]*Haboker*, "Electronic brain in Weizmann Institute," Feb. 23, 1955.

[169]Accad to Leviathan, personal communication, Oct. 24, 2010.

Fig. 3.9 Menachem Berl, 24-years old WEIZAC team member, feeding a problem to the machine (May 1, 1959). Courtesy of the Israel National Photo Collection (*Credit* Moshe Pridan)

Fig. 3.10 Anne Judith Donath, research assistant at DAM, working with WEIZAC (1962). Courtesy of WIS archives. *Photo Credit* Shlomo Ben Zvi. All rights reserved to WIS

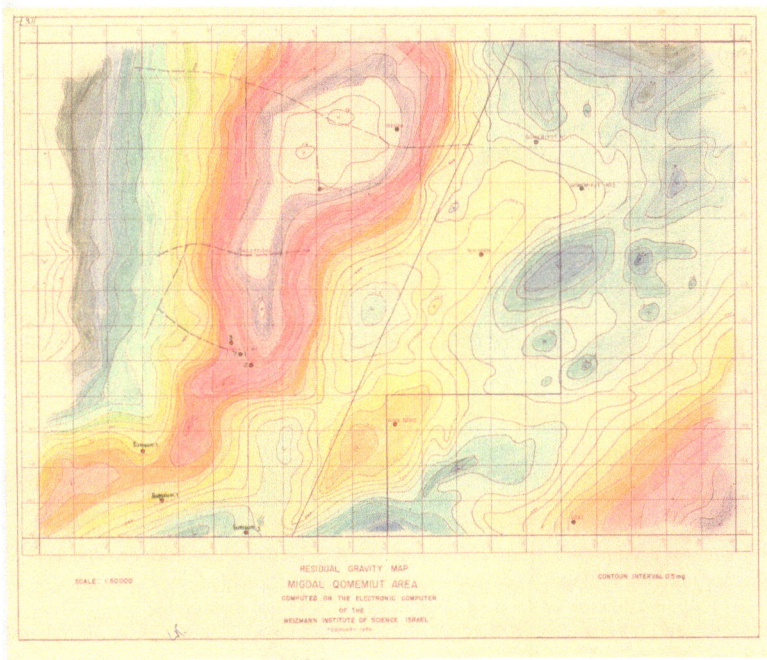

Fig. 3.11 Israel—Residual Gravity Map—Migdal Qomemiut Area of The Weizmann Institute of Science, Israel ISA-Collections-Maps-000sfqx. Courtesy of the State of Israel Archive

Fig. 3.12 WEIZAC computer. Courtesy of Weizmann Institute of Science

References

Alterman Z, Pekeris CL, Jarosch H (1959) Oscillations of the earth. Proc R Soc Lond A 252:88–95

Ariav G, Goodman S (1994) Israel: of swords and software plowshares. Commun ACM 37:17–21

Aspray W (2000a) The institute for advanced study computer: a case study in the application of concepts from the history of technology. In: Rojas R, Hashagen U (eds) The first computers. MIT Press, Cambridge, pp 179–194

Aspray W (2000b) Was early entry a competitive advantage? US universities that entered computing in the 1940s. IEEE Ann Hist Comput 22(3):42–87

Baal-Schem J (2007) The birth of a hi-tech society: first steps in electronics and computing in Israel. In: EUROCOM 2007. The international conference on "computer as a tool". IEEE, pp 2638–2640

Bacharach U (2009) By the power of the knowledge. The scientific corps of Israel [1948–1952]. N.D.D. Media Ltd.

Bigelow J (1980) Computer development at the Institute for Advanced Study Princeton. In: Metropolis N, Howlett J, Rota G-C (eds) A history of computing in the twentieth century. Academic Press Inc., London, pp 291–310

Blachman MN (1953) A survey of automatic digital computers, 1953. Office of Naval Research, Washington D.C.

Breznitz D (2002) Conceiving new industrial systems: the different emergence paths of the high-technology industry in Israel and Ireland. Shmuel Neaman Institute for Advanced Studies in Science and Technology. Technion—Israel Institute of Technology

Ceruzzi PE (2003) A history of modern computing. MIT Press, Cambridge

Corry L (ed) (2008) Fermat meets SWAC: Vandiver, the Lehmers, computers, and number theory. IEEE Ann Hist Comput 30(1):38–49

Davis PJ, Fraernkel AS (2007) Remembering Philip Rabinowitz. Not AMS 54(11):1502–1506

Eckert JP (1953) A survey of digital computer memory systems. Proc IRE 41(10):1393–1406

Eisner MD (1981) The old technologies. J Popul Cult, 157–163

Estrin G (1991) The WEIZAC years (1954–1963). IEEE Ann Hist Comput 13(4):317–339

Finerman A (1968) University Education in Computing Science. In: Conference on graduate academic and related research programs in computing science. Academic Press, New York

Franklin J (2001) John Blatt. Retrieved 14 Jan 2018, from web.maths.unsw.edu.au/~jim/blatt. html

Garb Y (2004) Constructing the trans-Israel highway's inevitability. Isr Stud 9(2):180–217

Hauptman HA (1990) History of X-ray crystallography. Struct Chem 1(6):617–620

Hughes TP (1983) Networks of power. Electrification in the Western society. 1880–1930. JHU Press, Baltimore

Jensen WB, Fenichel H, Orchin M (2011) Scientist in the service of Israel: the life and times of Ernst David Bergmann (1903–1975). Hebrew University Magnes Press

Kirsch RA (1998) SEAC and the start of image processing at the National Bureau of Standards. IEEE Ann Hist Comput 20(2):7–13

Krauss JD (1997) The origins and effects of Israel first digital computer: the story of the WEIZAC. Thesis presented to the Committee on History and Science in partial fulfilment of the requirements for the degree of Bachelor of Arts, Harvard University

Land F (2014) The story of LEO–the World's first business computer. Retrieved 27 Feb 2019, from https://warwick.ac.uk/services/library/mrc/explorefurther/digital/leo/story/

Lee C (2012) Obituary: Gerald Estrin, 90, UCLA computer science pioneer/UCLA Newsroom. Retrieved 17 Dec 2013, from http://newsroom.ucla.edu/portal/ucla/obituary-gerald-estrin-90-ucla-231609.aspx

Lehman M, Eshed R, Netter Z (1963) The checking of computer logic by simulation on computer. Comput J 6(2):154–162

Minker J (1977) Information storage and retrieval: a survey and functional description. ACM SIGIR Forum 12(2):12–108

Napper R (2000) The Manchester Mark 1 computers. In: Rojas R, Hashagen U (eds) The first computers: history and architecture. MIT, pp 365–378

Narasimhan R (1960) On the system and engineering design of the general purpose electronic digital computer at TIFR. In: Proceedings of the Indian academy of sciences-section A, vol 52, no 2. Springer, India, pp 47–57

Nebeker F (1993) Thelma Estrin, biomedical engineer: a pioneer of applied computing. Proc IEEE 81(10):1370–1382

Pekeris CL (1958) Ground state of two-electron atoms. Phys Rev 112(5):1649–1658

Pekeris CL (1964) A brief history of the department of applied mathematics. REHOVOT, p 3

Pomerene JH (1972) Historical perspectives on computers: components. In: Proceedings of the fall joint computer conference, part II, 5–7 Dec 1972. ACM, New York, pp 977–983

Pugh E (1984) Ferrite core memories that shaped an industry. IEEE Trans Magn 20(5):1499–1502

Puglisi JD (2009) Resolving the elegant architecture of the ribosome. Mol Cell 36(5):720–723

Racah G (1958) The use of the Weizac in theoretical spectroscopy. Bull Res Counc Isr F 8:1

Redmond KC, Smith TM (1980) Project whirlwind: the history of a pioneer computer. Digital Press, Bedford

Sanderson M, Croft BW (2012) The history of information retrieval research. In: Proceedings of the IEEE (Special Centennial Issue), vol 100, pp 1444–1451

Scheinerman ER, Simpson J (2001) The Fraenkel Festschrift volume. Electron J Comb 8(2)

Shahar A (2002) At the front of computing—Mamram: legacy of IDF computers center. Tel Aviv: Maarachot [Hebrew]

Stroke HH (1995) The physical review. The first hundred years: a selection of seminal papers and commentaries. American Institute of Physics, New York

Tedre M (2014) The science of computing—shaping a discipline. Chapman and Hall/CRC

Traub W, Yonath A (1969) Polymers of tripeptides as collagen models. IV. Structure analysis of poly(L-prolyl-glycyl-L-proline). J Mol Biol 43:461–477

Unna I (2000) The genesis of physics at the Hebrew University of Jerusalem. Phys Perspect 2:336–380

Weiselman IL, Tomash E (1991) Marks on paper: part 2. A historical survey of computer output printing. Ann Hist Comput 13(2):203–222

Wilkes MV (1951) The best way to design an automatic calculating machine. In: Proceedings of the Manchester University computer inaugural conference, pp 16–18

Xu J, Croft BW (1996) Query expansion using local and global document analysis. In: Proceedings of the 19th annual international ACM SIGIR conference on research and development in information retrieval. ACM, pp 4–11

Yonath A (2011) X-ray crystallography at the heart of life science. Curr Opin Struct Biol 21(5):622–626

Chapter 4
Concluding Remarks: WEIZAC as a Zionist Success Story

In the foregoing sections, we have discussed the story of WEIZAC as a success story located at the confluence of two important, essentially separate historical threads. One thread is that of the rise of the digital computer and of electronic computer technology in the years following World War II. The second thread is that of the creation of top-rated institutions of higher education and scientific research in Mandatory Palestine and, after 1948, in the recently created State of Israel, and in particular the early years of WIS as a budding center of cutting-edge scientific activity intent on pursuing its own agenda and in developing its own academic identity.

In this concluding chapter, we would like to summarize the main points discussed above, while stressing the central factors that contributed to the success of the WEIZAC project. A particularly illuminating, broader context within which this summary, and indeed the story as a whole, can be conveniently located is that of the role of science and technology as a decisive factor in processes of nation-building, with a specific focus on the post-WWII period. This topic has been the subject of recent historical research and it will be useful to briefly discuss some important points related to it before returning to WEIZAC.

In a special issue of the journal *OSIRIS* in 2009, for example, we find a collection of interesting case studies that explore the ways in which modern science and the nation-state grew up together in a process of fruitful dialogue and interaction. As the editors of the collection indicate, since the publication of Eric Hobsbawm and Terence Ranger's groundbreaking *The Invention of Tradition* (1983) much historical research devoted to questions of national identity had focused on the way in which national entities seek to anchor their legitimacy in a *past* that was often, as they argue, artificially fabricated. The focus on science and technology has helped introduce a parallel perspective to the discourse of national identity, which is rooted in modernity and which is "oriented toward the future rather than toward the past." A main example that highlights the gist of their presentation is that of India, which is always relevant and interesting as a point of reference for the Israeli case as well. When Jawaharlal Nehru in 1947 formed his first cabinet, in addition to assuming the prime ministership, he reserved for himself the ministry of science and technology. By combining both roles, Nehru believed, he would make law suited to the "scientific temper" of the citizens of a new and independent India (Harrison and Johnson 2009, 1).

In their introduction to the *OSIRIS* collection, Carol E. Harrison and Ann Johnson indicate the parallel between the processes recognizable in India, as well as in other contemporary arising nation-states, with that of the emergence of modern nation-states in the eighteenth-century Enlightenment. They laid the focus of their argumentation upon the common underlying belief that science would teach citizens to be sovereign and would draw them into a close relationship with their new state. In their own words:

> Frenchmen of the late eighteenth century built on the Enlightenment conviction that the institution of the state had to be brought into line with the basic rationality of the human mind; science was the means to make the French state conform most closely to human reason. Nehru, in contrast, understood himself to be following a model established by modernization in an array of European and American states. To be a nation- state, India needed science along with a middle class, an industrial economy, a national education system, representative government, a press—all elements taken from the example of Western modernization (Harrison and Johnson 2009, 2).

While as first prime minister of Israel, Ben-Gurion assumed also the position of minister of defense, rather than "of science and technology," as Nehru did, this kind of argumentation is clearly relevant to the case of Zionism as well. Zionist discourse was eager to present a scientific-oriented, utopian view of the future, starting from a past that involves the tradition of Jewish learning. Indeed, from its inception in the late nineteenth century, the Zionist ideology perceived the sciences, pure and applied, as central to its program of creating a new Jewish society in the Land of Israel. This view was rooted in the European Jewish Enlightenment, or "Haskalah," beginning in the late eighteenth century, which in turn was inspired by classical enlightenment values and was motivated by the wish to achieve a fuller integration of Jews into European society. Enlightened Jews, or "maskilim," advocated a revival of the Hebrew language and Hebrew literature and promoted the study of secular topics among Jews, including science. Accordingly, Hebrew books and journal dealing with scientific topics started to appear in places like Odessa, Berlin and Warsaw (Feiner 2004; Shavit and Reinharz 2011; Soffer 2004). In the mainstream version of secular Zionist ideology, modern science could provide relief for the Jews from their suffocating religion and, at the same time, would afford tools needed to recover their ancient land from its ruins. Sharing universal knowledge and the values of science with their European mates, the Jews would finally become both normal and self-determined.[1]

It was in this spirit that as early as 1897, at the First Zionist Congress in Basel, the Jewish mathematician and "maskil," Zvi Hermann Schapira (1840–1898), suggested the creation of a Jewish institution of higher learning in Palestine as a

---

[1]For a more detailed discussion see Corry and Golan (2010), which is the introduction to a special issue of the journal *Science in Context*, devoted to the history of science in the Israeli context.

main, immediate task for the movement that had just been created. He sketched a plan for an institution comprising research and teaching of secular disciplines in the framework of several schools of theology, theoretical sciences, technology, and agriculture. In his vision, this would become the cornerstone of the national revival of the Jewish people (Corry and Schappacher 2010, 441–443). The idea was elaborated into a more concrete plan only at the Fifth Zionist Congress of 1901, and Weizmann became its main promoter. The plan led, indeed, to the creation of the Hebrew University in Jerusalem and it thus connects in a rather concrete way to the story of WEIZAC as told above.

Another relevant collection of articles devoted to the topic of nation building and science, which is relevant to our discussion, appeared more recently in the journal *History and Technology*. Here the focus is on the opposite direction, namely, the way in which practitioners of science and technology joined the local political elites of the newly emerging nations in their pursuit of legitimacy to govern and to help achieve the "peoples' own quest for self-determination, national liberation, and prosperity in whatever forms those always-contested objectives might take" (Krige and Wang 2015, 171). Some of these practitioners were drawn into the corridors of political and diplomatic power and very often they found themselves at ease therein. John von Neumann has provided the most prominent—some would say notorious —example of this symbiotic relationship (Israel and Millán-Gasca 2009, 82–120). The story of Weizmann Institute in this regard is somewhat complex and many-sided. On the one hand we have the cases of Bergmann or Dostrovsky. As indicated above Bergmann became a central figure in the Israel Atomic Energy Commission. Dostrovsky, a physical chemist at WIS, served as first Director of Research of the same commission since 1953, and as Director-General from 1965 to 1971. Similarly, biophysics Ephraim Katchalski Katzir (1916–2009), was the first commander of HEMED, the Science Corps of Israel, and—emblematic of the symbiotic relationship between the nation and its leading scientific figures—in 1973 he became the fourth president of Israel. On the other hand, many of the scientists that initially followed Bergmann and his stances during the 1948 war later objected to concentrate on defense-oriented research. Thus on May 1954, a group of young nuclear physicists, led by Amos de Shalit (1926–1969), renounced their positions at the Ministry of Defense and moved to WIS with the intention to concentrate on purely scientific research.[2] WIS promised to make all research results available to the Atomic Energy Commission of Israel. However, Pekeris, in particular, strongly opposed to WIS submission to any censorship of the results of scientific work done at the Institute.[3] He was also reluctant to give computing time of WEIZAC to the Ministry of Defense personnel (Shahar 2002, 26).

One could bring additional examples of studies referring to these issues for the cases of various Latin-American nations (e.g. Corry (ed.) 2003; Corry (ed.) 2005;

---

[2]This episode became known as "The revolt of the physicists". See Jensen (2011, 195–19), Mardor (1981, 114–117), and the Hebrew version of Cohen (1998, especially 56–68).
[3]SCM held on Apr. 11, 1954 (CPA).

Medina 2011; Medina et al. (eds.) 2014), European countries (e.g. Schot et al. 2010), the Soviet Union,[4] and India as well as other Asian nations (e.g. Basset 2009, 15–32; Bassett 2016; Günnergun and Raina (eds.) 2011; Raina 2003). But one should not loose sight an important point made by the editors of the *History and Technology* collection, when they stress that the interaction between these practitioners and the political and military leadership was not limited to the obvious cases of "advanced weaponry, high technology, and large scientific establishments that came to define the symbolic and literal meanings of power in the nuclear age":

> Political elites who sought to take advantage of the new opportunities created by the changing world order to refashion the identities and trajectories of their nations turned to the transformative potential of science and technology to fill out the contours of imagined futures. … Cold war competition has loomed large in [existing accounts], given the significance of atomic weapons, nuclear reactors, rockets, and satellites as quintessential markers of security, modernity, and national prowess. … More recent scholarship on development has also drawn attention to agriculture, public health, scientific and technical aid, industrial policy, and myriad forms of social scientific investigation as modes of endeavor tied to cold war objectives (Krige and Wang 2015, 172).

In our case, there are several interesting examples of scientific-technological projects carried out in Mandatory Palestine and in the early years of the State of Israel, and in which the connection between Zionism and science and technology as main tool for nation-building plays a prominent role.

In her book *Healing the Land and the Nation: Malaria and the Zionist Project in Palestine*, for example, Sandra Sufian raises some fundamental questions about the relationship between technology, disease, and nationalism (Sufian 2007). Sufian surveys the fight against malaria in Mandatory Palestine, and explains its consequences over the topographical, demographic, and epidemiological landscape. This is an important topic because the story of the early Zionist settlers who suffered from malaria during their first years in Palestine, and that of the subsequent efforts for draining the swamps, have played a central role in the Zionist imaginary for generations (see also Davidovitch and Zalashik 2010). Sufian's study uncovers the ideological dimension attached to these efforts as a precondition for the transformation of the Diaspora Jews into "healthy persons," both physically and spiritually, as they return to their historical, biblical homeland. Medical science is thus presented as an important force for shaping a renewed national consciousness and as a bridge connecting the past and the future of the Jewish nation.

Also agricultural technology played a very important role in the efforts to settle the land and to establish economically viable Jewish colonies in Palestine and, like the fight against malaria, it has been a main topic in the Zionist imaginary. Nahum Karlinsky discusses the case of the citrus industry in his 2005 book *California Dreaming: Ideology, Society, and Technology in the Citrus Industry of Palestine, 1890–1939* (Karlinsky 2005). Of particular interest in Karlinsky's analysis is the focus on *ideologically conflicting views* within the Zionist movement, in relation

---

[4]See e.g., https://www.amazon.com/Stalins-Great-Science-Adventures-Physicists/dp/1860944191.

with this topic. Should the citrus industry develop along capitalist, private enterprise lines or along the communal lines advocated by mainstream Zionist socialists? Should the industry rely exclusively on Jewish manpower as ideologically promoted by Ben-Gurion and the Zionist Labor Movement, or could also Arab (cheaper) hired hands be employed? Karlinsky's analysis shows how these conflicting views affected the extent of, and the motivation for, embracing the adoption of new technologies and innovations in agriculture, with the common goal of establishing the Jewish national home. At the same time, these conflicting views obviously affected in various ways the mounting tensions between Arab and Jews in Palestine. But interestingly, while the Arab sector of the industry always tended to be more traditional and the Jewish sector more inclined to the introduction of new, advanced technologies, Karlinsky shows that capital obtained from Jewish land purchases ultimately brought about an expansion of the Arab citrus sector. In addition, Arab growers also learnt about modern techniques while working in Jewish orchards. Technological innovation, as shown in this example, could help deepening the gap between the two ethnic populations, but also at the same time bridging over the same gap.

A third interesting topic related to the question of technology and nation-building and that we would like to mention here concerns the story of electrification of Mandatory Palestine, and in particular the story of the construction of the electricity power network beginning in the 1920s. One study to have approached this question is Ronen Shamir's *Current Flow: The Electrification of Palestine* (Shamir 2013). A main point in Shamir's sociologically-oriented analysis is that while the electric grid, by its own nature, had the potential of spreading electricity all over the land, "regardless of the identities of communities and sectors" it ended up widening the gap between Arabs and Jews. Like Karlinsky in his analysis, Shamir also suggests that technological issues could sharpen ideological divides in the discourse of nation building within the Zionist camp. The moving force behind the electrification process of Palestine was Pinhas Rutenberg (1879–1942), a Russian-born engineer, youthful socialist turned ardent Zionist and capitalist entrepreneur. At both the ideological and practical level, he was much more closely associated with the revisionist leader Zabotinsky than with Ben Gurion and the Labor Movement. While Rutenberg's initiatives put forward the necessary infrastructure for the eventual creation of a Jewish state, they responded to a discourse and to motivations that were diametrically opposed to those underlying all other infrastructure projects promoted by the General Organization of Workers in Israel (the powerful "Histadrut") and the socialist-oriented leaders of the Yishuv.

A more recent publication dealing with the same issue is Fredrik Meiton's *Electrical Palestine: Capital and Technology from Empire to Nation* (2019).[5] Meiton builds on the metaphor of political power as inherently associated with electrical power, both of which circulated hand-in-hand over the grid that Rutenberg constructed. He specifically singles out electrification as a process that

---

[5]And see also Meiton (2016).

became a main driving force of the Zionist project of state-building, and associates the expansion of the electrical grid with the expansion of the Zionist settlement in Palestine. He thus focuses on the centrality of electricity and electrification as a crucial explanatory category that adds an important perspective to understand the processes that—in his analysis—led to the success of Zionism in establishing the Jewish state while, simultaneously, the Arab population in Palestine failed to establish their own.

A final point that seems relevant to consider in this brief overview concerns the question of Ben-Gurion's attitude towards the expected role of science and technology as part of the Zionist project. This complex issue has been studied by Ari Barell (2014). He characterized Ben-Gurion's attitude in terms of three interacting dimensions, that he also attributed, in various degrees, in the approach of other Zionist leaders as well. These are: (1) the practical dimension: science as a tool to develop the country; (2) the ideological-national dimension: science as a main component in the cultural revival of the Jewish people; (3) the epistemological-philosophical dimension: science as a most reliable method for clarifying the truth about the universe.

Worthy of particular attention is the fact that Ben-Gurion repeatedly expressed his interest in the WEIZAC project and indeed the entire idea of building of an automatic computer at WIS bore a direct relationship with the three above mentioned dimensions of his attitude towards science and technology. An interesting document that can be cited in this regard is a letter sent to Amos de Shalit on January 1957, where Ben Gurion expressed his philosophical concerns around the question of computers and the human mind. While doing so, however, Ben Gurion also went directly to the grain by stating that the real interest arisen by an electronic computer derives from the assistance it may provide to research on atomic physics and one may assume that he was particularly interested in the way that such a research would contribute to the defense interests of the Jewish state. Thus, Ben Gurion wrote:

> Our latest conversation at the apartment of my daughter Renana refuses to leave my mind. I am bothered by your opinion, as a physicist, concerning men's mind and consciousness. Can it be that your specialization and your expertise in the physical sciences overshadows your ability to acknowledge the superiority of the human spiritual and intellectual forces? If that is the case, then there is, in my opinion, a serious blunder in the professional education of the scientists.

> Do you even conceive that two machines can be built that will interchange letters similar to those of Spinoza about art, philosophy and science? Is it conceivable that a machine will travel around the world gathering detailed facts and eventually coming up with Darwin's theory?

> Can't you grasp the totally different essence of the processes that affect the human spirit (and which are undoubtedly connected to physical processes that affect the human body), whose nature is completely different from that of purely mechanical processes?

> Can you imagine a machine that would write up the Book of Job or Plato's "Symposium," or Einstein's theory of relativity?

The most sophisticated machine will perhaps be able to fulfill its builder's will, but there is almost no limit to the human mind and to the intellectual capabilities of man – and this is the only reason why I believe in the benefit of atomic research.[6]

Against the backdrop of this broader context and the comparative perspective that it provides, we can now move to our concluding assessment of the story of WEIZAC and of its success. The WEIZAC project was not, of course, one of infrastructure, in the straightforward, material sense of the word. Nor was it one that could directly contribute to issues of public health, housing, food provision, or education. It was not a project that would affect in any direct, relevant manner the relationship between Arabs and Jews citizens of the newly created state, even though, as most of the high-level scientific research and the high technology industry that would develop in Israel over the following decades, it inherently segregated any participation of Arabs. In retrospect, of course, we can see its fundamental contribution to setting the foundation for the computer industry and the computer culture at large in Israel, including health, defense and government issues. But as initially conceived of, WEIZAC was purely a tool intended for scientific research, and, as we saw above, one whose value even in that delimited realm, and under that relatively narrow definition, was strongly contested. Still, it is our claim, that it was only against the backdrop of the Zionist ethos and, more specifically, on the basis of the Zionist network of cooperation and support, particularly of the Jewish communities in the UK and the USA after the creation of the state, that the success of the project can be understood.

As we stressed repeatedly above, for Weizmann, developing a strong infrastructure of high-level, basic scientific research seen as end in its own sake, and as an activity of intrinsic value, was part of his Zionist vision for the Jewish national home. This was the view that underlies the creation of the Hebrew University in Jerusalem and later on, also the creation of the Sieff Institute. Compared to the Hebrew University, however, the institute at Rehovot promoted a more accommodating attitude towards applied science alongside pure research (and also it did not comprise faculties of humanities or social sciences, of course). Likewise, the leaders of WIS saw themselves as strongly committed to the overall effort of the Zionist movement seen as a project of nation-building, from a much more active perspective that many of the colleagues in Jerusalem.

Providing the necessary conditions for advanced scientific research would in Weizmann's view, enrich and deepen the creative forces, both spiritual and material, of the Jewish population in Palestine. In the long run, it would also set the basis for applied research that would improve the economic situation of the Yishuv and later on of the State of Israel. The creation of WIS would be the means for achieving that end while at the same time helping develop agriculture and industry,

---

[6]Ben Gurion to de Shalit, on Jan 13, 1957 (BGA). In a letter to Dr. Hans Kreitler, the founder of the psychology department at Tel-Aviv University, on 17 February 1963, Ben Gurion returned to this topic and confessed that he cannot stop thinking about "the mystery of human thought" (ISA-PMO-DirectorGeneralPMO-000m9nu).

based on the use of modern technologies. It would contribute to solving issues of nutrition and public health, while enhancing at the same time the ability to absorb educated Jews from all over the world. With the help of Bergmann and Weisgal, Weizmann was able to help materialize this vision through the creation of WIS.

Pekeris, in turn, strongly identified himself with these aims, which also fitted his own kind of Zionist conviction. Since he saw applied mathematics as a leading field of cutting-edge science, he was convinced that his own expertise and his plan for building an electronic automatic computer would easily match the institutional mission of WIS as promoted by its leaders, and would lead it to important achievements at the highest-level of international success. Seen in retrospect, he couldn't have done better in fulfilling his intended aims. Pekeris was very wise in setting for himself and for WIS goals that were both ambitious, as they aimed at building a state of the art computer in a world where there were very few thereof, and cautious, as he avoided the adoption of unproved technologies. His goal was the *implementation*, rather the *invention* of new technologies (Estrin 1991). The project was meant to enable scientific innovation and advancement, and at the same time to enhance the level of practical engineering expertise in the country.

The successful realization of a project of this kind required a full commitment of an entire organization, and Pekeris was definitely able to create this commitment and to put to work the entire organization, its financial resources, manpower and space, on behalf of WEIZAC. A perfect matching there arose between the tasks and goals of the project and those of the decision-makers at the different levels: at the individual level, at the project level, at the organizational level and even at the national level. Weisgal periodical reports on the scientific activities of WIS, in particular, make clear the extent to which the WEIZAC project was increasingly seen as a direct materialization of Weizmann's Zionist doctrine.[7] It is thus clear that Pekeris found the way to turn the electronic computer project into a vital part of "the organizational mission" of WIS, and this was crucial to the success of the project.

In this sense, it is interesting to compare the WEIZAC project to the one that served as its inspiration, namely, the IAS computer at Princeton. Here we have two institutions devoted to high-level scientific activity, with no teaching duties for its faculty members, so that they can fully concentrate on their research. The WEIZAC project, as already indicated, organically engaged into the organizational mission of WIS, though only gradually so. But, as William Aspray has stressed in his study of the IAS computer, there was no similar incorporation of the project into the "organizational mission" of the institute at Princeton (Aspray 2000). In his account of the IAS computer project, Georg Dyson vividly described his impression about the

---

[7]Weisgal, "Report to the Board of Directors of the American Committee and to Committees in Other Countries for the Weizmann Institute for the Period of 1944–1949" (1949) (WIA); Weisgal, "Report by the Chairman of the Executive Council for the Period from November 2, 1949 to Jun. 30, 1952 (draft) 195) (WIA), "Scientific Activity Report" 1953 (WIA).

computer project as being extraneous to the spirit of the institution, in the following words:

> Doing things with our hands and building dirty old equipment. That wasn't the institute... The coming of six engineers with their assortment of oscilloscopes, soldering irons, and shop machinery was something of a shock (Dyson 2012, 119).

And indeed, it is remarkable that in 1952, as von Neumann spent less and less of his time at IAS, eventually moving to UCLA,[8] the project was abandoned and, in 1957, the computer was sold to Princeton University where it was shut down due to the high costs of maintenance and operation. Still, its influence on the early computers scene, including the WEIZAC project, was clearly decisive. Pekeris and his team at WIS, contrary to von Neumann and his at IAS, was indeed able to turn WEIZAC into an organic part of the backbone of scientific activity at Rehovot.

Various nations emerged as newly created states in the post-WWII period, but the process of adoption of computer technology, and in particular the ability to build an electronic computer, arose only later in all of them than in Israel. This is in particular the case even in the three nations that, as mentioned above in Sect. 2.1, by the end of the twentieth century did figure prominently in the world of high technology and electronic computing: India, Ireland and Taiwan.[9] The early success of the WEIZAC project needs thus to be understood against the background of the peculiar historical conditions of the creation of WIS and of the prominent role accorded to science and technology within the Zionist project and then in the young State of Israel. But no less important for understanding this process is to stress, in a very focused manner, the role played by the Jewish diaspora at the time.

Indeed, it is peculiar to the case of the Zionist project that in various world-leading centers of scientific research there existed a strong presence of top-rated Jewish scientists. Many of them immigrated to Mandatory Palestine and later to Israel, either ideologically motivated or escaping the atrocities of Nazi Germany. They came mainly from Germany, Austria, Hungary and the UK, and later on from the USA, and they quickly filled up the faculties of the local institutions (the Hebrew University, the Technion, and WIS) turning them into high-quality research centers. This was the hotbed of scientific excellence within which the WEIZAC project emerged and was successfully carried out.

In this regard, it is particularly important to notice that the attitude of the Jewish diaspora towards the Zionist project underwent important changes in the 1940s, which had an immediate impact on the feasibility of scientific and projects of this kind. British Jewry, for one thing, underwent after 1936, the year of the Great Arab

---

[8]Von Neumann passed away in 1957, while still formally a member of the IAS.

[9]A detailed comparison of the role of science and technology in the respective nation-building processes would be well-beyond the scope of the present study. Still, it is worth mentioning here some existing studies that would be relevant to such a discussion. For the case of Ireland, see e.g., Fanning (2008), Fanning (2016), Harte (2007). For the case of India, see e.g., Aloysius (2000), Bassett (2009), Chaube (2012). For the case of Taiwan, see e.g., Chun (1994), Wachman (1994), Wang (2004), Yeh (2014).

Revolt, an important process of "conversion to Zionism" (Wendehorst 2012). The possibility of a Nazi invasion in 1940 and, more radically so, the revelations of the Holocaust in the spring of 1945 had a strong impact on a majority of British Jews. Facing the need to help the survivors and displaced European Jews, many came to believe that the most obvious answer needed to emerge by way of establishing a Jewish state in Palestine. As the British Mandate came to an end and the State of Israel was established in 1948, some British Jews immigrated and became Israeli citizens. But the preferred way to participate in the process of nation-building in the new Jewish state became, for British Jews as well as for Jews in other diasporas, donating money to Zionist funds (Wendehorst 2012, 253–260). Such donations had always been fundamental to the activities of the Zionist movement, of course, but now they became more meaningful and ubiquitous among members of the community. The Sieff Institute was created back in 1934 with the financial support of British Jews mainly from Manchester. Such contribution continued to be crucial for the further development of WIS, as well as of all other institutions of scientific research and higher education in Palestine and then in the newly created state.

Likewise interesting, and perhaps of greater impact for the WEIZAC project, was the case of the American Jewish community. Aware of the imminent and systematic destruction of European Jewry the previously existing resistance to the Zionist project in Palestine diminished considerably and was gradually replaced by unequivocal support for the creation of a Jewish state. Like in the case of British Jewry, this support was sometimes translated into a personal decision to immigrate, but more often it was manifest by means of other forms of material support.[10] The American Committee for WIS, established on October 1944, was only one important manifestation of this trend. Underlying the activities of the committee was a belief in a set of universal values that, in their view, should be common to the American Jewish community and that of Palestine. The leading role of science and technology was a central aspect of this set of values. Members of the committee were proud of mobilizing their energies around the enormous success of a world-class research center, bearing the name of the emblematic Zionist leader who happened to be a famous scientist as well. And on the other side of the equation were local activists of the caliber of Weisgal and Bergmann who knew very well how to capitalize on the attitude and vision of the American Committee (Cohen 2016, 115–155; Feldestein 2006, 1–40).

But not only the advantage of American Zionist philanthropy became crucial in terms of providing financial support to WIS and to projects like WEIZAC. Rather, there was also a qualitative advantage resulting from the network of connections with leading academic centers such as Princeton, Harvard, and MIT. This translated

---

[10]Another important way to support the state was by settling in Israel for a delimited period of time in connection with a specific professional project, for instance in the educational, medical or academic realm. This was indeed the case of the Estrins, who explicitly defined themselves as non-Zionists. In their own words: "Thelma and I are Jewish but we had never been exposed to Zionist thinking; in fact we held idealistic humanist dreams of "one world" with all nations working together for the good of mankind" (Estrin 1991).

into significant acts of technology transfer that explain, as we saw above in detail, the actual success of the project: manuals and advanced technical literature, in-site professional training of key players, and even purchase of specific components. In this sense, the specific circumstances associated with the attitudes of British and American Jewry in the wake of the Holocaust, against the background of existing, if incipient, scientific institutions in Palestine and the presence of a score of talented scientists who could materially carry out the project turned WEIZAC into an emblematic success-story of the Zionist project.

The project of building an electronic automatic computer stood in the mid-1950s at the forefront of global technology and it gradually became fundamental for developing top-quality theoretical and applied science at WIS. Several decades later, this kind of cutting-edge scientific research (at WIS, as well as at other academic institutions in the country), alongside a well-developed computer culture and a prosperous computer-based industry have become central to the image of the Israeli society as a one-of-a-kind Start-Up Nation. In the first decade of the State of Israel, when electronics was in its infancy, this computer project could be successfully carried out as a result of a truly peculiar combination of factors. This comprised the organizational commitment of WIS to a nation-building mission, a well-conceived national as well as scientific vision on the side of its founders, and, above all, Pekeris's unique blend of scientific talents and commitment to his own Zionist goals.

References

Aloysius G (2000) Nationalism without a nation in India. Oxford University Press, Delhi

Aspray W (2000) The institute for advanced study computer: a case study in the application of concepts from the history of technology. In: Rojas R, Hashagen U (eds) The first computers. MIT Press, Cambridge, MA, pp 179–194

Barell A (2014) Engineer-king: David Ben-Gurion, science and nation building. Ben Gurion University Press, Sede Boker, Israel

Bassett R (2009) MIT-trained swadeshis: MIT and Indian nationalism, 1880–1947. OSIRIS 24:212–230

Bassett R (2016) The technological Indian [e-book]. Harvard University Press, Cambridge, MA

Chaube SK (2012) Politics of nation building in India. Gyan Publishing House, Delhi

Chun A (1994) From nationalism to nationalizing: cultural imagination and state formation in postwar Taiwan. Aust J Chin Aff 31:49–69

Cohen A (1998) Israel and the bomb. Columbia University Press, New York

Cohen U (2016) From political rejection to scientific renewal: Weizmann and the establishment of Daniel Sieff research institute in Rehovot, 1931–1934. In: Cohen U, Chazan M (eds) Weizmann the leader of Zionism. Zalman Shazar Center for Jewish History, Jerusalem, pp 380–383 [Hebrew]

Corry L (ed) (2003) Studies on science in Latin America, Special Issue of Estudios Interdisciplinarios de América Latina y el Caribe 14(1)

Corry L (ed) (2005) Science in the Latin American context. Special issue of Sci Context 18(2)

Corry L, Golan T (2010) Introduction. Sci Context 23(4) ("Science in an Israeli Context: Case Studies"):393–399

Corry L, Schappacher N (2010) Zionist internationalism through number theory: Edmund Landau at the opening of the Hebrew University in 1925. Sci Context 23(4):427–471

Davidovitch N, Zalashik R (2010) Pasteur in Palestine: the politics of the laboratory. Sci Context 23(4):401–425

Dyson G (2012) Turing's cathedral. The origin of the digital computer. Pantheon Books, New York

Estrin G (1991) The WEIZAC years (1954–1963). IEEE Ann Hist Comput 13(4):317–339

Fanning B (2008) The quest for modern Ireland: the battle of ideas. 1912–1986. Irish Academic Press, Newbridge

Fanning B (2016) Irish adventures in nation-building. Manchester University Press, Manchester

Feiner S (2004) The Jewish enlightenment. University of Pennsylvania Press, Philadelphia

Feldestein A (2006) Ben-Gurion, Zionism and American Jewry. 1948–1963. Routledge, London

Günergun F, Raina D (eds) (2011) Science between Europe and Asia: historical studies on the transmission, adoption and adaptation of knowledge. Springer, New York

Harrison CE, Johnson A (2009) Introduction: science and national identity. OSIRIS 24(1):1–14

Harte L (2007) Modern Irish autobiography: self, nation and society. Palgrave Macmillan, New York

Hobsbawm E, Ranger T (eds) (1983) The invention of tradition. Cambridge University Press, Cambridge

Israel G, Gasca AM (2009) The world as a mathematical game. John von Neumann and twentieth century science. Translated by Ian McGilvay. Birkhäuser Verlag, Basel/Boston

Jensen WB, Fenichel H, Orchin M (2011) Scientist in the service of Israel: The life and times of Ernst David Bergmann (1903–1975). Hebrew University Magnes Press

Karlinsky N (2005) California dreaming: ideology, society, and technology in the citrus industry of Palestine, 1890–1939. State University of New York Press, Albany

Krige J, Wang J (2015) Nation, knowledge, and imagined futures: science, technology, and nation-building, Post–1945. Hist. Technol. 31(3):171–179

Mardor MM (1981) RAFAEL: On the Path of Research and Development for Israel's Security

Medina E (2011) Cybernetic revolutionaries: technology and politics in Allende's Chile. MIT Press, Cambridge, MA

Medina E, da Costa Marques I, Holmes Christina (eds) (2014) Beyond imported magic: science, technology and society in Latin America. MIT Press, Cambridge, MA

Meiton F (2016) Electrifying Jaffa: boundary-work and the origins of the Arab-Israeli conflict. Past Present 231(1):201–236

Meiton F (2019) Electrical Palestine: capital and technology from empire to nation. University of California Press, Oakland

Raina D (2003) Images and contexts: the historiography of science and modernity in India. Oxford University Press, New Delhi

Schot J, Rip S, Lintsen H (2010) Technology and the making of the Netherlands. The age of contested modernization, 1890–1970. MIT Press, Cambridge, MA

Shahar A (2002) At the front of computing—Mamram: Legacy of IDF computers center. Maarachot, Tel Aviv [Hebrew]

Shamir R (2013) Current flow: the electrification of Palestine. Stanford University Press

Shavit Y, Reinharz J (2011) The scientific god. Popular science in eastern Europe in the second half of the 19th century: between knowledge and a new image of the university. Hakibbutz Hameuchad, Tel Aviv [Hebrew]

Soffer O (2004) The case of the Hebrew press. From the traditional model of discourse to the modern model. Writ Commun 21(2):141–170

Sufian SM (2007) Healing the land and the nation: malaria and the Zionist project in Palestine, 1920–1947. University of Chicago Press, Chicago

Wachman AM (1994) Taiwan: national identity and democratization. M.E. Sharpe, Armonk, NY

Wang H (2004) National culture and its discontents: the politics of heritage and language in Taiwan, 1949–2003. Comp Stud Soc Hist 46(4):786–815

Wendehorst SEC (2012) British jewry, Zionism, and the Jewish state, 1936–1956. Oxford University Press, Oxford

Yeh H (2014) A sacred bastion? A nation in itself? An economic partner of rising China? Three waves of nation building in Taiwan after 1949. Stud Ethn Natl 14(1):207–228

Correction to: WEIZAC: An Israeli Pioneering Adventure in Electronic Computing (1945–1963)

Correction to:
L. Corry and R. Leviathan, *WEIZAC: An Israeli Pioneering Adventure in Electronic Computing (1945–1963)*, SpringerBriefs in History of Science and Technology https://doi.org/10.1007/978-3-030-25734-7

In the original version of the book, the following corrections have been made:

In the book front matter, the acknowledgement was missing. This has now been added.

Chapters 2 and 3 were inadvertently published with missed author corrections, which have now been incorporated.

In the book back matter, belated corrections received to remove bibliography section have been updated.

The updated version of the book can be found at
https://doi.org/10.1007/978-3-030-25734-7

© The Author(s), under exclusive licence to Springer Nature Switzerland AG 2019   C1
L. Corry and R. Leviathan, *WEIZAC: An Israeli Pioneering Adventure in Electronic Computing (1945–1963)*, SpringerBriefs in History of Science and Technology, https://doi.org/10.1007/978-3-030-25734-7_5

Appendices[1]

Appendix A: Scientific Publications Based on Calculations with Early Computers

Computer	Physical review papers between 1954 and 1963	Number of citation	Average number of citation per paper
UNIVAC I (University of Cal. Livermore, California)	27	1100	40
MANIAC	33	1500	50
WEIZAC	14	1885	134

Appendix B: Scientific Publications Based on Calculations with WEIZAC (Total: 65)

Publication	Institution or department involved
Abraham, C., & Aharoni, A. (1960). Linear decrease in the magnetocrystalline anisotropy. Physical Review, 120(5), 1576	WIS, Department of Electronics
Abrahamson, P., Ben-Arieh, J., Yekutieli, G., & Alexander, G. (1959). Interaction of 4.2 Gev π-mesons with nuclear emulsion. *Il Nuovo Cimento (1955–1965)*, 12(1), 27–37	WIS, Department of Physics

(continued)

---

[1]The data presented in the tables here below was compiled using GOOGLE search engine and the Physical Review on line archive. The search was conducted in 2014.

© The Author(s), under exclusive licence to Springer Nature Switzerland AG 2019
L. Corry and R. Leviathan, *WEIZAC: An Israeli Pioneering Adventure in Electronic Computing (1945–1963)*, SpringerBriefs in History of Science and Technology, https://doi.org/10.1007/978-3-030-25734-7

(continued)

Publication	Institution or department involved
Accad, Y., & Pekeris, C. L. (1964). The K2 tide in oceans bounded by meridians and parallels. *Proc. R. Soc. Lond. A*, 278(1372), 110–128	WIS, DAM
Aharoni, A., & Frei, E. H. (1960). On the Resolving Time and Flipping Time of Magnetoresistive Flip-Flops. *Proceedings of the IRE*, 48(8), 1436–1448	WIS, Department of Electronics & Stanford Research Institute, Menlo Park, Calif
Aharoni, A., & Shtrikman, S. (1958). Magnetization curve of the infinite cylinder. *Physical Review*, 109(5), 1522	WIS, Department of Electronics
Aharoni, A., Frei, E. H., & Shtrikman, S. (1959). Theoretical approach to the asymmetrical magnetization curve. *Journal of Applied Physics*, 30(12), 1956–1961	WIS, Department of Electronics
Alterman, Z., & Kornfeld, P. (1963). Propagation of a pulse within a sphere. *The Journal of the Acoustical Society of America*, 35(10), 1649–1662	WIS, DAM
Alterman, Z., Frankowski, K., & Pekeris, C. L. (1962). Eigenvalues and Eigenfunctions of the Linearized Boltzmann Collision Operator for a Maxwell Gas and for a Gas of Rigid Spheres. *The Astrophysical Journal Supplement Series*, 7, 291	WIS, DAM
Alterman, Z., Jarosch, H., & Pekeris, C. L. (1959). Oscillations of the Earth. *Proc. R. Soc. Lond. A*, 252(1268), 80–95	WIS, DAM
Birk, M., Goldring, G., & Wolfson, Y. (1959). Lifetimes of 2+ rotational states. *Physical Review*, 116(3), 730	WIS, Department of Nuclear Physics
Bosendorff, S., & Eisenberg, Y. (1958). Multiple scattering measurements in nuclear emulsions. *Il Nuovo Cimento (1955–1965)*, 7(1), 23–38	WIS, Department of Nuclear Physics
Bregman, J., Hirshfeld, F. L., Rabinovich, D., & Schmidt, G. M. J. (1965). The crystal structure of [18] annulene, I. X-ray study. *Acta Crystallographica*, 19(2), 227–234	WIS, Department of X-ray crystallography
Chavet, I., & Levanone, I. (1962). Absorption of Evaporated ZnS in the Ultra-violet. *Proceedings of the Physical Society*, 80(5), 1105	Israel Atomic Energy Commission

(continued)

(continued)

Publication	Institution or department involved
Coppens, P., & Schmidt, G. M. J. (1965). The crystal structure of the metastable (β) modification of p-nitrophenol. *Acta Crystallographica*, 18(4), 654–663	WIS, Department of X-ray crystallography
Coppens, P., & Schmidt, G. M. J. (1964). X-ray diffraction analysis of o-nitrobenzaldehydes. *Acta Crystallographica*, 17(3), 222–228	WIS, Department of X-ray crystallography
Davis, P., & Rabinowitz, P. (1958). Additional abscissas and weights for Gaussian quadratures of high order. Values for n = 64, 80, and 96. *J. Res. Nat. Bur. Standards*, 60(6), 613–614	National Bureau of Standards & WIS, DAM
Dostrovsky, I., Fraenkel, Z., & Friedlander, G. (1959). Monte Carlo calculations of nuclear evaporation processes. III. Applications to low-energy reactions. *Physical Review*, 116(3), 683	Brookhaven National Laboratory, Upton, New York & WIS, Department of Isotope Research
Dostrovsky, I., Fraenkel, Z., & Winsberg, L. (1960). Monte carlo calculations of nuclear evaporation processes. iv. spectra of neutrons and charged particles from nuclear reactions. *Physical Review*, 118(3), 781	Lawrence Radiation Laboratory, University of California, Berkeley, California & WIS, Department of Isotope Research
Dostrovsky, I., Rabinowitz, P., & Bivins, R. (1958). Monte Carlo calculations of high-energy nuclear interactions. I. Systematics of nuclear evaporation. *Physical Review*, 111(6), 1659	Los Alamos Scientific Laboratory, University of California, Los Alamos, New Mexico & WIS, Department of Isotope Research
Fraenkel, Z., Raviv, A., & Klein, W. (1963). Plate by plate calculations of multicomponent distillation columns using difference equations—I: Ideal cascades and constant flow columns operating at total reflux. *Chemical Engineering Science*, 18 (11), 697–709	Israel Atomic Energy Commission & WIS, Department of Isotope Research
Gillis, J. (1958). An application of electronic computing to X-ray crystallography. *Acta Crystallographica*, 11(12), 833–834	WIS, DAM
Gillis, J. (1961). Stability of a column of rotating viscous liquid. In *Mathematical Proceedings of the Cambridge Philosophical Society* 57(1), 152–159. Cambridge University Press	WIS, DAM
Goldring, G., & Vager, Z. (1962). Gyromagnetic Ratios of the 2+ States in Even Tungsten Isotopes. *Physical Review*, 127(3), 929	WIS, Department of Nuclear Physics

(continued)

(continued)

Publication	Institution or department involved
Goren, R., & Monselise, S. P. (1965). Inter-relations of hesperidin, some other natural components and certain enzyme systems in developing Shamouti orange fruits. *Journal of Horticultural Science*, 40 (2), 83–99	Hebrew University of Jerusalem, Department of Citriculture, Faculty of Agriculture, Rehovot
Gorodetzky, S., Scheibling, F., Armbruster, R., Benenson, W., Chevallier, P., Mennrath, P., ... & Goldring, G. (1963). Radiative Corrections in the Angular Correlation of Monopole Pairs from O 16 at Small Angles. *Physical Review*, 131(3), 1219	Institut de Recherches Nocleaires and Centre de Recherches Nucleaires du Centre National de la Recherche Scientifique Strasbourg, France & WIS DAM
Grover, N. B., Goldstein, Y., & Many, A. (1961). Improved representation of calculated surface mobilities in semiconductors. I. Minority carriers. *Journal of Applied Physics*, 32(12), 2538–2539	Hebrew University of Jerusalem, Department of Physics
Hordvik, A. (1966). The Crystal and Molecular Structure of Rhodan Hydrate. *ACTA CHEMICA SCANDINAVICA*, 20(3), 754–770	Chemical Institute, University of Bergen, Bergen, Norway
Jaffe, J. H., Goldring, H., & Oppenheim, U. (1959). Infrared dispersion of absorbing liquids by critical angle refractometry. *JOSA*, 49(12), 1199–1202	WIS, Section of Infrared Spectroscopy.
Jaffe, J. H., Meiron, J., & Jacobi, N. (1962). Instrumental Corrections in Measurements of Gas Dispersion. A Three-Slit Problem. *JOSA*, 52(1), 8–14	WIS, Section of Infrared Spectroscopy.
Kogan, A. (1960). A Practical Method for Calculation of Pressure Distribution on Airfoils in Supersonic Rotational Flow. Weizmann Science Press of Israel 1960	Technion, Israel Institute of Technology
Kogan A., & Bester, A.A. (1960). On Supersonic Flow Past Thick Airfoils. *Journal of the Aerospace Sciences*, 27(7), 504–508	Technion, Israel Institute of Technology
Lehman, M., Eshed, R., & Netter, Z. (1963). The checking of computer logic by simulation on a computer. *The Computer Journal*, 6(2), 154–162	Israel Ministry of Defense, Scientific Department
Longman, I. M. (1960). On the utility of Newton's method for computing complex roots of equations. *Mathematics of Computation*, 187–189	WIS, DAM

(continued)

(continued)

Publication	Institution or department involved
Low, W., & Rosengarten, G. (1964). The optical spectrum and ground-state splitting of Mn2+ and Fe3+ Ions in the crystal field of cubic symmetry. *Journal of Molecular Spectroscopy*, 12(4), 319–346	Hebrew University of Jerusalem, Department of Physics
Maor, U., & Yekutieli, G. (1960). The excited nucleon's model of meson production. *Il Nuovo Cimento (1955–1965)*, 17(1), 45–67	WIS, Department of Nuclear Physics
Maradudin, A. A., & Weiss, G. H. (1959). Auxiliary Integrals for Lattice Sums. *The Journal of Chemical Physics*, 31(5), 1433–1433	Institute for Fluid Dynamics and Applied Mathematics, University of Maryland, College Park Maryland & WIS DAM
Meiron, J. (1959). Automatic lens design by the least squares method. *JOSA*, 49(3), 293–298	Israel Ministry of Defense, Scientific Department
Meiron, J., & Loebenstein, H. M. (1957). Automatic correction of residual aberrations. *JOSA*, 47(12), 1104–1109	Israel Ministry of Defense, Scientific Department
Meiron, J., & Volinez, G. (1960). Parabolic approximation method for automatic lens design. *JOSA*, 50(3), 207–211	Israel Ministry of Defense, Scientific Department
Mendlowitz, H. (1963). Transition Array for d^3 \to d^2p: VANADIUM III. *The Astrophysical Journal*, 138, 1277	National Bureau of Standards, Washington DC
Nahmani, G., & Davids, N. (1962). Scabbing of steel by plane stress waves. *International Journal of Mechanical Sciences*, 4(1), 73–81	Israel Ministry of Defense, Scientific Department & University of Pennsylvania
Pauncz, R., & Ginsburg, D. (1960). Conformational analysis of alicyclic compounds—I: Considerations of molecular geometry and energy in medium and large rings. *Tetrahedron*, 9(1–2), 40–52	Technion, Israel Institute of Technology, Department of Chemistry
Pekeris, C. L. (1963). Final Report on an Investigation of the Three Body Problem in Atomic Physics	Air Force Office of Scientific Research through the European Office, Aerospace Research, United States Air Force
Pekeris, C. L. (1960). Propagation of seismic pulses in layered liquids and solids. In *International symposium on stress wave propagation in materials, New York, Interscience Publishers, Inc*	WIS, DAM
Pekeris, C. L. (1962). Excited S states of helium. *Physical Review*, 127(2), 509	WIS, DAM
Pekeris, C. L. (1958). Geophysics, pure and applied. *Geophysical Journal of the Royal Astronomical Society*, 1(3), 257–262	WIS, DAM

(continued)

(continued)

Publication	Institution or department involved
Pekeris, C. L. (1958). Ground state of two-electron atoms. *Physical Review*, 112(5), 1649	WIS, DAM
Pekeris, C. L. (1959). 1^1 S and 2^3 S States of Helium. *Physical Review*, 115(5), 1216	WIS, DAM
Pekeris, C. L., & Alterman, Z. (1957). Solution of the Boltzmann-Hilbert integral equation II. The coefficients of viscosity and heat conduction. *Proceedings of the National Academy of Sciences*, 43(11), 998–1007	WIS, DAM
Pekeris, C. L., & Lifson, H. (1957). Motion of the surface of a uniform elastic half-space produced by a buried pulse. *The Journal of the Acoustical Society of America*, 29(11), 1233–1238	WIS, DAM
Pekeris, C. L., & Longman, I. M. (1958). The motion of the surface of a uniform elastic half-space produced by a buried torque-pulse. *Geophysical Journal International*, 1(2), 146–153	WIS, DAM
Pekeris, C. L., Jarosch, H., & Alterman, Z. (1959). Dynamical theory of the bodily tide of the earth. In *Report on 'Third International Symposium on Earth Tides', Trieste* (pp. 17–18)	WIS, DAM
Pekeris, C. L., Longman, I. M., & Lifson, H. (1959). Application of ray theory to the problem of long-range propagation of explosive sound in a layered liquid. *Bulletin of the Seismological Society of America*, 49 (3), 247–250	WIS, DAM
Rabinovich, D., & Schmidt, G. M. J. (1964). 387. Topochemistry. Part V. The crystal structure of 2, 5-dimethyl–1, 4-benzoquinone. *Journal of the Chemical Society (Resumed)*, 2030–2040	WIS, Department of X-ray crystallography and DAM
Rabinowitz, P. (1960). Abscissas and weights for Lobatto quadrature of high order. *Mathematics of Computation*, 47–52	WIS, DAM
Rabinowitz, P. (1960). Abscissas and weights for Lobatto quadrature of high order. *Mathematics of Computation*, 47–52	WIS, DAM
Rabinowitz, P., & Weiss, G. (1960). Tables for the Calculation of Lattice Sums. *The Journal of Chemical Physics*, 33(4), 1272–1273	WIS, DAM

(continued)

(continued)

Publication	Institution or department involved
Racah, G. (1959). Use of the Weizac in theoretical spectroscopy. *Bull. Res. Counc. of Israel F*, 8, 1–14	Hebrew University of Jerusalem, Department of Physics
Racah, G., & Shadmi, Y. (1959). The configurations (3d+ 4s) n in the second spectra of the iron group. *Bull. Res. Counc. of Israel F*, 8, 15–46	Hebrew University of Jerusalem, Department of Physics
Racah, G., & Spector, N. (1960). the configurations $3d'' 4p$ in the second spectra of the iron group. *Bulletin: Mathematics and physics*, 9, 75	Hebrew University of Jerusalem, Department of Physics
Sanders, J. B. (1961). Spectroscopic calculation of energy levels of some tin isotopes. *Nuclear Physics*, 23, 305–311	Institute Lorentz, University of Leyden, Leyden, The Netherlands
Schiff, B., & Pekeris, C. L. (1964). f Values for Transitions between the 1^1 S, 2^1 S, and 2^3 S, and the 2^1 P, 2^3 P, 3^1 P, and 3^3 P States in Helium. *Physical Review*, 134(3A), A638	WIS, DAM
Pekeris, C. L., Schiff, B., & Lifson, H. (1962). Fine Structure of the 2^3 P and 3^3 P States of Helium. *Physical Review*, 126(3), 1057	WIS, DAM
Talmi, I., & Unna, I. (1960). Energy levels and configuration interaction in Zr90 and related nuclei. *Nuclear Physics*, 19, 225–242	WIS, Department of Nuclear Physics

Appendix C: Timeline of Events

	History of computing	Weizmann institute of science
1934		April: The Daniel Sieff Institute for Chemical Research is officially opened
1943	September: the relay interpolator (Model II) of Bell Labs is operational in in New York. Colossus became operational in England	September: Pekeris gets to know the automatic digital computer
1944		October: the American committee for WIS is established
1945	June: first draft of a report on the EDVAC	
1946	February: the ENIAC is completed	

(continued)

(continued)

	History of computing	Weizmann institute of science
	Opening of IAS computer project. The Moore School Lectures on design of electronic digital computers	Pekeris joins the planning committee of WIS. Pekeris joins the meteorological project at IAS under the direction of von Neumann
1947		A decision is taken to support the construction of an automatic computer at WIS
1948		May: the declaration of independence of the state of Israel
1948		December: Pekeris arrives at Rehovot and takes the position of the head of DAM
1949	June: stored-program computer, EDSAC, of Cambridge, becomes operational	November: WIS is formally inaugurated
1950		Frei joins the IAS computer project
1951	UNIVAC is delivered to Census Bureau	
1952	IAS computer is operational	January: the scientific committee of WIS decides to support the construction of an electronic computer in Rehovot
1953		December: the Estrins arrive in Rehovot
1954		July: Work on the computer at WIS begins
1954	April: first commercial magnetic core memory is installed on the JOHNNIAC computer	
1954		November: the electronics lab at WIS is inaugurated
1955		March: WEIZAC is officially presented
1955		April: the Estrins return to the US
1955		Summer: first programming course by Rabinowitz
1955		September: WEIZAC becomes operational
1956		September: Magnetic core memory is installed in WEIZAC
1957	February: John von Neumann passes away	
1958		"Ground state of two-electron atoms" is published
1961		July: the TRANSAC computer arrives in Israel and becomes operational with the IDF
1963		December: WEIZAC is turned off

Index